PATHWAYS TO THE UNIVERSE

PATHWAYS TO THE UNIVERSE

Francis Graham-Smith

Astronomer Royal
Director of Nuffield Radio Astronomy Laboratories,
Jodrell Bank
Professor of Radio Astronomy,
University of Manchester

Bernard Lovell

Emeritus Professor of Radio Astronomy,
University of Manchester

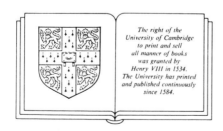

*The right of the
University of Cambridge
to print and sell
all manner of books
was granted by
Henry VIII in 1534.
The University has printed
and published continuously
since 1584.*

CAMBRIDGE UNIVERSITY PRESS

Cambridge

New York   New Rochelle   Melbourne   Sydney

Published by the Press Syndicate of the University of Cambridge
The Pitt Building, Trumpington Street, Cambridge CB2 1RP
32 East 57th Street, New York, NY 10022, USA
10 Stamford Road. Oakleigh, Melbourne 3166, Australia

First published 1988

Printed in Great Britain by Scotprint Ltd, Musselburgh, Scotland

British Library Cataloguing in Publication Data
Graham-Smith, Francis
Pathways to the universe.
1. Astronomy
I. Title   II. Lovell, Bernard, Sir
520   QB43.2

Library of Congress Cataloguing in Publication Data
Graham-Smith, F. (Francis), 1923–
Pathways to the universe
Bibliography: p.
Includes index.
1. Astronomy.   I. Lovell, Bernard, Sir, 1913–
II. Title.
QB43.2.G72 1988   520   86–24517

ISBN 0 521 32004 6

DS

Contents

Acknowledgements

We are greatly indebted to Dr Jacqueline Mitton for her help and advice and particularly for assistance with the illustrations and their captions. We also wish to thank Dr Anne Cohen for reading and correcting the early drafts of the text.

FG-S
BL

Preface

There have been two major revolutions in the science of astronomy. The first, in the 16th and 17th centuries, was associated with Copernicus and Galileo, the realisation that the Earth is not stationary at the centre of the Universe and the first use of an optical telescope. The second occurred in the middle of this century when observations of the Universe reached beyond the limits of optical observations and embraced the new techniques of radio, X-ray, ultraviolet, infrared and gamma-ray astronomy. Astronomy also became a part of space research, with the most spectacular results. We have been privileged to see this transformation during our research careers; we took part in the development of the new techniques and in the new understanding that they have given us. We have also attempted to share with others the thrill of discovery and our sense of wonder about the Universe; these are the reasons for this book.

An important centre of these new developments has been at Jodrell Bank. Public interest became so great that in 1966 it was necessary to make a special exhibition for visitors. The Science Centre now contains a planetarium and an exhibition of modern astronomy, where each year we introduce 100 000 visitors to our new science. The visitors are very demanding: amateurs and professionals alike require simplicity, clarity and precision in our displays. More than half are schoolchildren, and for many of these the visit is their first introduction to modern scientific research. We have modelled this book on the lessons we have learned from our visitors.

The topics of each chapter are intended to stand alone. Each starts from experience that we can all share, such as a simple look at the night sky with binoculars, and each ends by taking the reader close to the frontiers of modern research. Such rapid progress involves drastic selection, but there are many other books which provide more detail on amateur observations, on astrophysics and on cosmology.

Astronomy is one of the most approachable of all scientific subjects; it can form an introduction to scientific education and understanding. We invite our readers to share with us our excitement and pleasure in understanding a little of the Universe that is now being revealed to us.

FG-S
BL
Jodrell Bank

1

How we became astronomers

What is the nature of the Universe and what is our place in it? We all seek the answers to these questions, but the answers often seem to be hidden in mathematics or astrophysical theory. We, the authors of this book, think that science can be simple, and we like it to be accessible to all. Our simple approach to astronomy, which we want to communicate in this book, follows from our own unconventional introductions to the subject. Both of us started as scientists, trained in physics but not in astronomy. Let us start by introducing ourselves.

During World War II, we were both involved in the development of radar. Before that time, one of us (BL) had already started a research project at Manchester, on cosmic rays. Our experiences in radar led us both away from the familiar university research topics and into the first days of radio astronomy, one (FG-S) at Cambridge and the other (BL) at Jodrell Bank. Here, in the early post-war years, we saw the start of a revolution in astronomy that has continued for half a century. Previously, the optical telescope was the only way to explore the Universe, simply because it was believed that the light emitted by the stars and galaxies was the sole means through which we could gain information about the Universe. The revolution in astronomy started with the discovery that radio waves could be received on Earth and that they, like light waves, brought us information about the Sun, the stars, the galaxies, and the Universe as a whole. This was the beginning of radio astronomy, and it was soon followed by other new astronomies: X-ray, infrared, ultraviolet, and gamma-ray astronomies, all using their own techniques for exploring the sky.

On the day that World War II began in September 1939, Bernard Lovell was already at an operational radar station, looking at short-lived echoes on the radar screen. The operator said that these had no connection with aircraft but were reflections from ionisation in the atmosphere. This generated the idea that radar might be a useful technique for the detection and study of the showers of atomic particles that develop in the atmosphere after the impact of very high energy particles from outer space, the cosmic rays. Six years later, after the end of the war, came the opportunity to test the idea. Trailers of borrowed radar equipment arrived in a field at Jodrell

Opposite Sir Bernard Lovell (right) and Sir Francis Graham-Smith (left) in front of the 250-foot Lovell Telescope at Jodrell Bank during the 30th anniversary celebrations in 1987.

Bank and a search began for echoes from the cosmic ray showers. The experiment was intended to be simple and quick, and the field was borrowed from the Botanical Department of the University of Manchester for only two weeks. The surprise result was that echoes were obtained not from cosmic ray showers but from meteor trails and, consequently, the two weeks have now extended to over 40 years! During this time the whole new subject of radio astronomy has developed and the radio astronomy observatory has completely overwhelmed the botanical trial grounds. Ironically, the cosmic ray shower echoes have never been detected, and there is good reason for believing that they are too weak to have been detected by the simple radar systems used in that early work. The meteor trails, however, turned out to be very exciting.

As a cosmic ray physicist, Bernard Lovell then faced an unfamiliar problem. The radar echoes not only told him the direction of the meteor trails but they continued to appear in the day-time when no visual meteor observations could be made. How were the directions of these trails to be described? Were the daytime meteors interesting? The answers to these questions came not from books or from the lecture room but by sitting in the open through many nights by the side of an expert meteor observer, J.P.M. Prentice. He was a solicitor by profession but had a profound knowledge of the sky. So started the story of meteor radar astronomy, which we set out in a

Bernard Lovell (centre of front row) and the radio astronomy research group at Jodrell Bank in 1951 pictured in front of what they called 'the searchlight aerial'.

later chapter, and so started the astronomical education of one of the authors.

Graham Smith similarly returned to university research after the wartime radar work, joining Martin Ryle at the Cavendish Laboratory, Cambridge. Ryle was already following up some war-time observations, not of radar reflections but of radio waves emitted by the Sun (Chapter 12). By a fortunate coincidence, the largest sunspot for many years occurred at this time, in July 1946, and Cambridge radio astronomy grew from observations of its very powerful radio emission. Soon afterwards, the same techniques were used to detect radio waves coming from other parts of the sky. An American communications engineer, Karl Jansky, as long ago as 1932, had found that the whole of the Milky Way was a powerful source of radio waves; his observations were followed up only by a lone enthusiast, Grote Reber, who constructed a parabolic radio telescope in his back garden in Wheaton, Illinois. Again, following wartime observations in Britain, another pioneer, J.S. Hey, showed that part of this cosmic radio signal came from a discrete source in the constellation of Cygnus. This was to be the main target of our research at Cambridge.

When Ryle and Smith set up their simple radio telescope, in a field behind the house where they both lived, they left it for 24 hours to scan the sky as the Earth rotated, recording the radio signal on a paper chart recorder. When they looked at the chart on the follow-ing morning, they were astonished to see the trace of not one but two powerful radio sources on the recording. Suddenly they found themselves at the forefront of astronomy but, like Bernard Lovell at Manchester, neither knew his way about the sky sufficiently well to specify the location of either source. It took a day before they could say that this new radio source 'appeared to be in a constellation called Cassiopeia'.

Only a few years later this naive but correct statement was super-seded by a precise determination of the position of the new radio source, known as Cassiopeia A. Optical astronomers could then be asked to help and Cassiopeia A was soon identified with the

F. Graham-Smith with Martin Ryle at Cambridge in the early 1950s.

remains of a stellar explosion in the Milky Way, i.e. a supernova (Chapter 16). The first radio source, Cygnus A, was identified with an entirely different object, a very distant galaxy (Chapter 19).

Both authors, therefore, found themselves with a new key to the Universe but almost completely lacking in basic astronomical information. Nevertheless, our scientific training soon took us and our colleagues into one of the richest fields of modern astronomy, and we soon found ourselves members of the international community of astronomers.

After many years as a radio astronomer at Cambridge and at Jodrell Bank, Graham Smith moved into optical astronomy as Director of the Royal Greenwich Observatory. This switch illustrates the close link between radio and optical astronomy. The main task at the time was to build the new optical telescope at La Palma, in the Canary Islands. The techniques of using radio and optical telescopes are sufficiently similar that experience in one field could be transferred directly to the other. Bernard Lovell also found himself involved with astronomy on a broad front, taking the Chair on committees that were overseeing the rapid advances in all branches of UK astronomy during the 1950s and 1960s. Both authors have faced the need to understand modern astronomical research, starting from an almost complete lack of formal training. Our intention in this book is to carry the reader with us along our own path of understanding and instruction.

Grote Reber, the pioneer of radio astronomy, photographed in 1960 at the National Radio Astronomy Observatory in Green Bank, West Virginia, with the radio telescope he built in 1937.

A photograph taken on 10th October 1933 of Karl Jansky at Bell Telephone Laboratories pointing to the Milky Way on a rough chart of the sky. This was the area of the sky from which cosmic radio signals were first detected. Jansky was attempting to pinpoint the source of noise interfering with radiotelephone services.

Visitors to Jodrell Bank

The 250-foot radio telescope at Jodrell Bank, now known as the Lovell Telescope, first operated in October 1957. It was equipped with a radar that was originally used in the meteor radar research, but that was now aimed at detecting echoes from the carrier rocket of the first man-made Earth satellite, Sputnik 1. There was an immediate rush of visitors to Jodrell Bank, all wanting to see the telescope in action. There was great excitement, too, when on 13 September 1959 the telescope was used to record the arrival at the Moon of Lunik II, the first successful space probe. At this time, there were very few tracking stations for space probes and the telescope was in great demand. On 11 March 1960 it was used to transmit a vital command to the American probe Pioneer V and continued to track it out to a distance of 36 million kilometres.

The constant publicity about the association of the telescope with the Soviet and American space projects led to large numbers of requests from individuals and groups who wanted to visit Jodrell. The staff soon found it impossible to deal with even the most deserving cases. Bernard Lovell suggested to the University that we

The Science Centre at Jodrell Bank stands close to the 250-foot Lovell Telescope. Thousands of people visit the exhibition and planetarium every year.

should make some provision for satisfying this legitimate public interest in the way that other observatories, particularly in America, had done. There was some doubt whether the demand would justify the expense and it was suggested that public reaction might be tested by erecting a marquee and holding two weeks of open days during which lectures would be given. This experiment was tried in the summer of 1964 and on the first Sunday afternoon the queues of cars waiting to gain access stretched over two miles from the entrance. We charged a modest entrance fee to cover the cost of the marquee and 35 000 people came in those two weeks. Even so, the University was not convinced that there would be a continuing interest, although there was a prediction that 50 000 visitors could be expected each year. The experiment was repeated during the summer of 1965 – with a similar result. Some of the staff had volunteered to collect the entrance fee, but they were overwhelmed and eventually we placed a series of buckets into which our visitors could throw their money.

After these experiences the University agreed that we should erect a building that would serve the needs of the visitors on a regular basis. This building, originally known as the Concourse,

Young visitors at the Jodrell Bank Science Centre explore the properties of black holes as they experiment with the 'gravity hollow'.

was officially opened by His Grace the Duke of Devonshire, the Chancellor of the University, on May 3rd 1966. A planetarium was added in 1971. The Concourse has now expanded into a major exhibition of modern science, which forms, with the planetarium, a Science Centre catering for upwards of 90 000 visitors a year, about half of whom are children. The millionth visitor to the Science Centre was welcomed in March 1978.

Here is a magnificent opportunity for presenting science to a wide public, especially the many schoolchildren who visit us. How should we go about it, and what science should we present to complement the science curriculum in school?

The main message we seek to present in the Jodrell Science Centre is that science is both a vital activity in the modern world and a fascinating and exciting pursuit. We also seek to show that scientists are ordinary people who enjoy their work. Ask a young schoolchild to draw a scientist, and you will probably receive a caricature of a wild, remote figure in the midst of unpleasant chemical apparatus or sinister nuclear devices. Astronomy, which has none of these connotations, provides an attractive and important introduction to science for anyone. Many examples of the basic laws of physics can be found in this book: the vast scales of time, space, energy, and density inside and between the stars extend these laws to conditions unobtainable on Earth. There is also a message about modern technology: much of the equipment and techniques used in astronomical research is related to the technology that society now considers essential. There is much to say and many different kinds of people to address.

We have both spent long hours thinking about the presentation of science to the public. We find it essential to start always at the most basic level but, at the same time, we want to share our sense of excitement in the most modern of astronomical research techniques and results. Our experience in the Science Centre led to the idea of this book. We decided that each chapter should take an individual theme from a starting point that anyone can recognise. The experienced observer of the night sky will have some advantage, since he or she will recognise some of the individual stars and galaxies that provide our examples. Modern astronomy does, however, take us into less familiar ground; we must introduce techniques such as X-ray and gamma-ray astronomy, and concepts such as nucleogenesis and black holes. We seek to show how complex and strange ideas grow naturally out of the simple and familiar.

We share with other astronomers a deep conviction of the value and importance of our subject. Hundreds of students have studied at our radio observatory and have made their mark on the world in as many different ways. For a scientific training it is an excellent discipline. Our motivation is, however, quite straightforward. Faced with the most fundamental questions about the Universe and provided with the means at least to approach the answers to them, we feel that we have no choice but to proceed as hard as we can. And having enjoyed the privilege of pursuing research in a new field, we are now impelled to tell the story to the best of our abilities.

2

Finding your way about the stars

The first step in astronomy is easy. Choose a clear night and a comfortable place to lie or sit and look at the stars for half an hour or so. A few bright stars can be seen immediately, unless city lights are too dazzling; within five minutes the eyes become dark-adapted and we can see hundreds or even thousands more. At first glance the stars are fixed points of light, but after a while the whole pattern is seen to be moving steadily from east to west. If we watch from night to night we may see some of the brighter points of light, the ones that are the planets, move among the background of the stars; we may also see the Moon, in a different position each night. If we observe through the year we can see a gradual change, in an annual cycle.

Let us leave aside for a moment our binoculars and telescopes, which tell us so much of the nature of the celestial objects, and simply describe the sky as it has been seen by innumerable people over many thousands of years. The pattern of stars appears as if painted on the inside of a sphere, rotating majestically overhead and carrying the pattern from east to west. We see the pattern from the centre of this celestial sphere.

The steady east–west motion of the stars is circular, centred on the two poles of the celestial sphere. In the northern sky, there happens to be a bright star, the Pole Star, near the centre of rotation. A group of bright stars, part of the constellation of the Great Bear (Ursa Major), called the Big Dipper, the Plough, or Charles' Wain, provides a signpost to the north celestial pole. The two stars known as the Pointers lead the eye towards the Pole Star at all times. Through the night, the line of the Pointers acts like the hand of a clock, rotating around the north pole. Two quick looks at the sky an hour or two apart show this clearly. A camera left with the shutter open and directed towards the pole will photograph the circular tracks of the stars. Another conspicuous constellation, Orion, lies close to the equator of the celestial sphere; for observers near the terrestrial equator, this constellation passes nearly overhead. In the south, there is no bright star at the pole; the famous Southern Cross is some way away and rotates round the pole like the Pointers do in the north.

Stretching round the whole sky like a diffuse band of light is the Milky Way. This is unmistakable and spectacular from a dark site; sadly, there are many city dwellers who have never even noticed it because of the brightness of city lighting. In the northern hemisphere, the Milky Way is seen at its best late in a summer evening,

when it stretches from high in the sky down to the southern horizon. The Milky Way is our Galaxy, a disc-like cloud of stars on a scale vast even compared with the distances of the individual bright stars of the familiar constellations. Our Sun, with its planet Earth, is one of the hundred thousand million stars of this Galaxy. Other galaxies, of similar size, can be seen at great distances from the Milky Way. Southern hemisphere observers can see two of these distant galaxies easily without the help of binoculars; they are the two Magellanic Clouds, which at first sight look like detached patches of the Milky Way. In the northern hemisphere, observers have to look very hard to see the only 'extragalactic nebula' that is bright enough to be visible by the naked eye. This is the Andromeda Nebula, a galaxy very like our own in size and shape.

How are we to find our way about the sky, so that we can pick out quickly the various interesting stars and galaxies? If we see a meteor or a comet, which may appear at any point against the background of the stars, how are we to describe its position? We need first to be familiar with the main constellations.

Mapping the sky – the constellations

Although we cannot see the stars in the daytime, we can give the position of the Sun at any time against the background of the constellations. In the course of a year, the Sun completes one circuit of the celestial sphere, following a path known as the ecliptic. The ecliptic is not far from the celestial equator, being inclined to it at an angle of 23°. The constellations along the ecliptic are well known: they are the twelve houses of the Sun, the 'zodiac' of astrology. Isaac Watts helps us to remember them thus:

> The Ram the Bull the Heavenly Twins
> And next the Crab, the Lion shines
> The Virgin and the Scales;
> The Scorpion, Archer and the Goat,
> The Man that pours the water out
> The fish with glittering scales.

As Sir John Herschel said in 1849, 'the constellations seem to have been almost purposely named and delineated to cause as much confusion as possible'. There are 88 constellations officially recognised by astronomers though not many people could name them all, let alone place them accurately in the sky. Nevertheless, there are some easily recognisable features and some bright stars, both on and away from the ecliptic, which provide a simple framework for finding our way about the sky. In the north, our simple guides are the Plough (the Big Dipper), Orion and the three bright stars we call the Summer Triangle. Cassiopeia, a conspicuous W shape in the Milky Way, can be found from the Pointers of the Plough by extending their line past the Pole Star an equal distance and slightly to the right. Cassiopeia is close enough to the pole to be seen rotat-

NORTHERN HEMISPHERE

The brightest constellations in the
northern half of the celestial sphere.

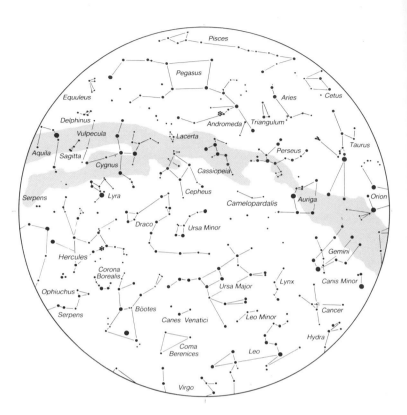

SOUTHERN HEMISPHERE

The main constellations in the southern
part of the sky.

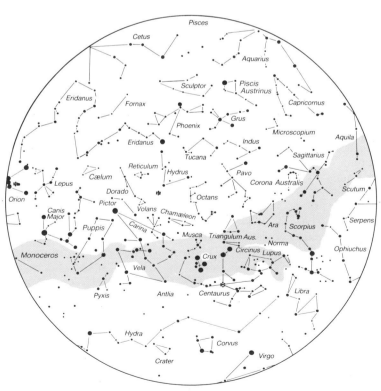

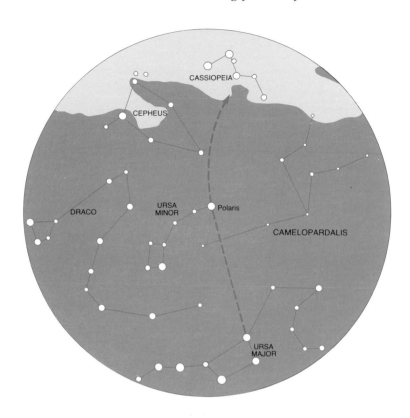

As seen from much of Europe and North America, the stars in this chart are circumpolar, that is, they never set but sweep out circles around the north celestial pole.

ing round it, like the Pointers. Extending the line from the Pole Star beyond Cassiopeia we reach the square of Pegasus.

Our line from the Pole Star crossed the Milky Way in Cassiopeia. If we now follow the Milky Way towards the right, we move first into Cygnus (the Swan), a simple cross whose short tail is marked by the star Deneb, pointing towards Cassiopeia. The long neck of the Swan points along the Milky Way towards the less striking constellations of Vulpecula, Aquila, Scutum and Sagittarius, towards the centre of our Galaxy. The northern Milky Way constellations are best seen in the summer and autumn. A useful guide to the summer sky is provided by the three stars Deneb in Cygnus, Altair in Aquila and Vega in Lyra. Of these, Vega is the most westerly and the brightest.

In spring a different triangle is a useful guide; it is made up of Arcturus (in Boötes), Regulus (in Leo) and Spica (in Virgo). These stars are on the opposite side of the Plough from the Pole Star. In the northern winter the most conspicuous pointer is Orion (the Hunter), easily recognised by his belt of three bright stars, his head the bright red star Betelgeuse in the north and his foot Rigel in the south. Orion's belt points left (for northern hemisphere observers) towards Sirius (in Canis Major) and right to Aldebaran (in Taurus). Beyond Aldebaran lies the conspicuous open cluster, the Pleiades.

Betelgeuse can also be used as a pointer to Castor and Pollux, the Heavenly Twins (Gemini), and Capella in Auriga. Orion is south of the Milky Way; from this region the Milky Way runs through Auriga and Perseus back to Cassiopeia.

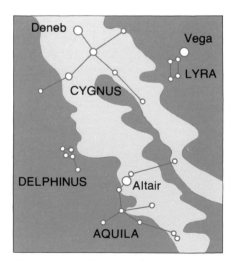

For northern observers, the 'summer triangle' of the stars Deneb, Altair and Vega are a useful reference frame for finding other constellations.

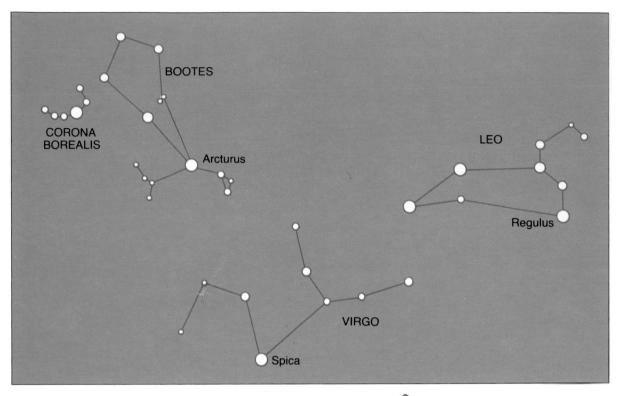

The constellations Boötes, Leo and Virgo feature prominently in the evening sky for northern observers in spring.

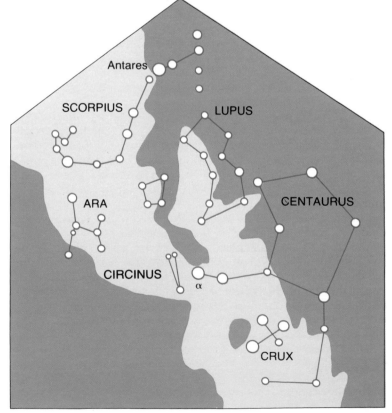

The Milky Way and the bright constellations through which it passes are the most prominent features of the southern sky.

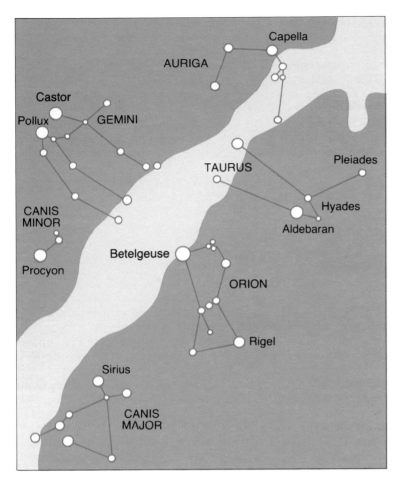

Orion is possibly the most readily identifiable constellation of all. Located near the celestial equator, the bright constellations around Orion are visible to both northern and southern observers.

Observers in the southern hemisphere can use Orion and Sirius as starting points for observing in winter evenings. Scorpius, with the bright star Antares, is another good starting point in the summer. Antares is outside the Milky Way but, without question, the best feature of the southern sky is the Milky Way itself, extending from Scorpius through Norma, Lupus, Circinus and Centaurus to Crux, the Southern Cross. The Cross, a brilliant group of four stars outlining a kite, is at the southernmost point of the Milky Way, which then stretches through Carina, Vela, Puppis and Monoceros to the region of Orion, opposite Scorpius.

For most of us such a quick and sketchy guide is sufficient. We can glance up at the sky with some general familiarity but without the detailed knowledge of the dedicated observer hunting for comets or variable stars. The next stage is for the keen observer to work with published star charts; several newspapers produce very useful charts every month, showing also the positions of the planets and the Moon. After that, a more detailed star atlas will be needed.

The planets

All the planets are seen close to the ecliptic, in the zodiacal constellations through which the Sun passes. Mars, Jupiter and Saturn can appear as bright objects at any point along the ecliptic; Uranus can sometimes just be seen by the naked eye if you know exactly where to look, but Neptune and Pluto require both a finding chart and a telescope. Venus is very often seen as a brilliant object in the morning or evening sky; it is never more than 47° from the Sun. Mercury is harder to see as it is always very close to the Sun and often lost in the twilit sky.

A particular challenge for a keen observer is to follow Venus through a cycle of 1½ years. As its apparent distance from the Sun changes, so does its brilliance and its appearance; it is a crescent when it is most brilliant. Mars and the other outer planets always show as more or less full discs.

Constellations and coordinates

Just as a geographical position can be precisely described by a latitude and longitude, the position of a star can be described by coordinates on the celestial sphere. Declination, like latitude, is measured in degrees, positive north of the equator and negative for south. Right ascension, corresponding to longitude, is measured in hours, minutes and seconds, the complete circle of 360° corresponding to 24 hours. The two poles of this system are fixed by the rotation axis of the Earth; they are the points in the sky directly over the Earth's north and south poles.

One might reasonably expect the axis of the Earth to stay pointing in a fixed direction in relation to the stars; in fact it does not. The axis, and with it the whole system of celestial coordinates, moves slowly and steadily across the sky, tracing out a circle over a period of 26 000 years. This slow movement of the pole, which is called precession, makes very little difference to the celestial coordinates of a star over a period of a few years, but it makes a big difference over recorded history. Three thousand years ago, for example, the north celestial pole was in the constellation of Draco and near the star α Draconis; it is now close to the star Polaris, the Pole Star, and slowly moving closer to it.

As the pole moves, so does the part of the sky visible from any fixed place on the Earth. This gives us a fascinating clue to the origin of the names of the constellations that astronomers use today. We know that 48 of the constellations were named in prehistoric times, since Ptolemy listed them in 150 BC and they were evidently in use very much earlier. A large area of the sky was left unnamed presumably because it was below the horizon for the civilisation that named the constellations. The area left unnamed was in the south but it is not centred on the present-day south celestial pole; it is centred instead on the position of the pole in 2500 BC. This gives the date when the constellations must have been defined and named. It also gives the latitude of the civilisation which named them; the answer corresponds with the Babylonian and the Mediterranean regions in which there were flourishing civilisations in 2500 BC. Most probably the constellations were important to a seafaring nation, so it seems likely that the people of the Minoan civilisation on the Mediterranean islands of Crete and Thera were the original inventors of these constellations.

Star positions and proper motions

The constellations deceive us into thinking that their member stars are truly related to one another in space. They are in fact only the chance patterns seen from Earth of stars that are often at very different distances from us and totally independent of one another. If we could watch the skies for long enough we would see not only the movement of the pole due to precession but also changes in the actual patterns of the constellations due to unrelated movements of the individual stars. The familiar Plough, or Big Dipper, after 100 000 years would show five of its members in nearly the same configuration, but two obviously out of place. These are the nearer members of the constellation Ursa Major, whose motion through space is most obvious to us. Their movement across the sky is known as 'proper motion'.

The existence of proper motion means that catalogues of accurate star positions soon become out of date. The measurement of accurate star positions is very important to astronomers, and the proper motions are a key to understanding the structure of our Galaxy. The traditional observational work of positional astronomy, or astrometry, is still carried on as a background activity in several observatories. One such is at the Roque de los Muchachos Observatory on La Palma in the Canary Islands, where astrometry is done by an automatic instrument (called a transit circle) operated jointly by Danish, British and Spanish astronomers.

A transit circle is a telescope that looks only at the meridian, the north–south line through the sky over the observatory, so it is mounted in a single east–west bearing. At each observation of a star crossing the meridian the elevation above the horizon and the time it crosses the meridian are recorded. A special series of observations are made of the Sun whose motion defines the plane of the ecliptic, and of the planets. Observations over periods of some years provide measurements of the proper motions of the stars and of the precession of the whole coordinate system.

Fortunately for the everyday observer, these proper motions are very small. Their existence was discovered by Edmond Halley in 1718, when he compared his own observations with those recorded by Ptolemy dating back by about 1800 years. In that long interval, only a few stars had obviously moved, although some of the brightest had moved by as much as half a degree.

The brightness of the stars

The brightness of a star is called its 'magnitude'; stars visible by the unaided eye range from about zero magnitude (the brightest) to magnitude five or six (the faintest). The scale of magnitudes was devised by Hipparchos in 150 BC, and it has been found to be so convenient that it has been extended to the faintest stars and galaxies observable by modern telescopes, which may be as faint as magnitude 25. Space telescopes, like the Hubble Space Telescope, should extend the range to magnitude 27. The magnitude scale is a scale of ratios: one magnitude is a ratio of about 2.5 in brightness, and five magnitudes corresponds to a ratio of 100.

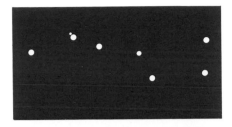

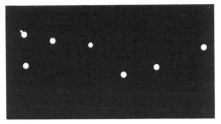

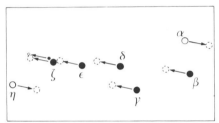

In 100 000 years time, the familiar shape of the Plough will have changed noticeably because of the individual movements of the stars through space. The apparent change in position in the sky is called proper motion.

Comets and meteors

If our half-hour look at the sky happens to be during the first half of August, it is very likely that some bright shooting stars, or meteors, will be seen. These will be part of a shower known as the Perseids: all of their tracks, projected back across the sky, meet in the constellation Perseus. More will be observed after midnight than before.

Meteors can be observed at any time of the year, but there are more to see at the time of the various meteor showers. The Perseids in August and the Geminids in mid-December provide the most regular and reliable displays. All meteors are part of the solar system, travelling in orbits round the Sun. The Aquarids and the Orionids are two showers in the same highly elliptical orbit, and the two shower periods (May and October) are the times when the Earth, moving in its own orbit, intersects the ellipse. Significantly, Halley's Comet is moving in the same elliptical orbit. Comets and meteors are closely linked. There are observed instances of comets breaking up into small pieces, and it is obvious that comet tails consist of material evaporated or broken from the surface of the comet head. Material from Comet Halley has spread round the elliptical orbit, and is now seen as the Aquarid and the Orionid meteor showers.

Comets and meteors consist of the original material that condensed to form the whole solar system. In the Sun, and on the surfaces of the planets, this material has been altered, but in the comets it remains in its most primitive form. It was therefore a most exciting scientific challenge to take a close look at the head of a comet; in March 1986 five spacecraft flew close by Comet Halley and sent back some remarkable photographs. The next stage will be a rendezvous in which a spacecraft can take a sample of this primitive material and bring it back to Earth.

Some of this original solar system material can be found on the Earth's surface, without any need for a spacecraft. A meteorite is a meteor that is sufficiently massive to penetrate the atmosphere and fall to the ground without burning up. Small meteorites, a few millimetres across, have been found in large numbers in certain places in the Antarctic; large meteorites, falling to Earth with an explosive force, may weigh several tonnes and, fortunately, are rare.

Now our quick glance at the sky is complete, and we have a background for our survey of modern astronomy.

3

Optical telescopes

A good pair of binoculars provides such an excellent means for observing the heavens that it is worthwhile considering in some detail how they work. To start with, it is not essential to have a 'pair'; one of the authors (FGS) uses an 8 × 30 monocular, in effect a small telescope, in preference to binoculars, as in his experience only one eye is usually in action when binoculars are in use. But why choose 8 × 30, either for bird-watching or for star gazing?

Binoculars act rather differently from cameras, which focus distant points on to a photographic film. The eye itself acts as a camera, focussing light on to the retina, and so do large astronomical telescopes, which focus starlight on to a detector such as a photographic plate, a photometer, or a spectrometer. Binoculars act as aids to the eye in two ways, helping by enlarging the angular spread of a distant scene and also by collecting more light in an objective lens which is larger than the pupil of the eye. These two functions are specified in the 8 × 30 label: the magnification is 8 times, and the objective lens is 30 mm in diameter.

Greater magnification is usually associated with larger diameter; for example, 12 × 60 is used by determined bird-watchers, although an instrument of this size is heavy and harder to hold steady. Higher magnification needs a steadier hand, and for astronomical use it is essential to have a tripod or other mounting. A larger objective lens is, however, useful for collecting more light. Why are magnification and objective diameter closely linked, so that we cannot choose, say, a 6 × 60 or a 10 × 20?

Try looking at the bright sky through your binoculars held 50 cm away from your eyes. You will see a round patch of light in the centre of each eyepiece. This patch should be about the size of the pupil of the eye, about 4 or 5 mm in diameter; if it is, then no light is wasted, and the eye is being used efficiently. The diameter of the patch (the exit pupil) is found by dividing the objective diameter by the magnification, so that 8 × 30 gives a diameter of 30 ÷ 8 or 3.75 mm, which is nearly optimum. The patch of light should, incidentally, be uniformly bright and clean; this is a test for the quality of the optical design and construction. Now bring the binoculars to the normal viewing position, and observe how the bright patch expands to fill the eyepiece lens. If you use spectacles, it is a great advantage to keep them on when using binoculars; this is only possible in a good instrument where the eyepiece is filled without having to come too close to the eye. This is known as having a large 'eye relief', and it involves a larger diameter eyepiece lens. Obviously the field of view in binoculars is less than that of the unaided

The path taken by a ray of light in one half of a pair of binoculars. The two prisms have the dual purpose of 'folding' the light path so that the binoculars are a manageable size and making the image the right way up.

eye. At best the ratio is approximately equal to the magnification; the more expensive instruments which are designed for this are usually labelled 'wide angle'. Binoculars with a much smaller field of view can be made more compact, but are not very useful.

As with all kinds of telescopes, the light gathering power depends on the area of the objective. If you are using binoculars for stargazing, then the bigger the better. But to see extended objects, such as a nebulosity, or for bird-watching in poor light, the gain from the larger lens is not so great. It is then more important to have good quality optics, and all glass surfaces coated to reduce reflection. In good binoculars the brightness of a uniform sky should be almost the same as it is for the naked eye. The lenses must, of course, also be colour corrected, avoiding 'chromatic aberration', which may be seen as coloured borders at the edges of objects near the edge of the field of view.

Why do we have 'prismatic' binoculars? The prisms, which may be the pair used in the standard angled instrument or the compact roof prism of the straight telescope, invert the image, which would otherwise appear reversed. The zig-zag light path also reduces the length. But there are low-power binoculars, known as 'opera glasses', which do not contain prisms. Understanding the difference takes us into a very interesting historical digression.

An engraving in the book *Operere di Galileo Galilei*, published in Bologna in 1656. Galileo is shown presenting his telescope to the muses and pointing out a heliocentric solar system.

Galileo, Kepler and Newton

Galileo, the first man to observe the sky through a telescope, learnt of the idea of combining two lenses to make a spyglass from Holland and from Paris. He immediately worked out the principle of combining a convex objective with a concave eyepiece, made his own lenses and showed that it worked. The same principle is used today in opera glasses. It seems remarkable that spectacle makers had not discovered the principle of this Galilean telescope much earlier. However, the lenses needed are unusual; a weak convex is to be combined with a strong concave, and if they are tried in contact the combination actually diminishes rather than magnifies. The distance apart of the lenses must be the long focal length of the convex, minus the short focal length of the concave lens.

The problem with the Galilean telescope is its restricted field of view. Kepler introduced the use of a convex lens for the eyepiece, as in our modern binoculars, and obtained a great improvement at the cost of inverting the image, as seen from the crossing of the rays in the figure. The obvious cure for the inversion is to add a further convex lens, and this is seen in various early examples: the problem is that the field of view is again reduced. If inversion is not tolerable, then the prisms of the binoculars are the best solution.

The next problem for the early designers was image quality. A large magnification needs a large ratio of focal lengths between the objective and the eyepiece, and eyepieces with very short focal lengths are both hard to make and small in diameter. So the objective focal length was made longer. Hevelius in Danzig made a telescope 45 m long, which was so clumsy it can scarcely have been useful. Huygens separated the objective and eyepiece lens completely, mounting the objective on a tower and the eyepiece on a stand up to 65 m away. This aerial telescope was useful: with one such Huygens made the first discovery of a satellite of Saturn. In

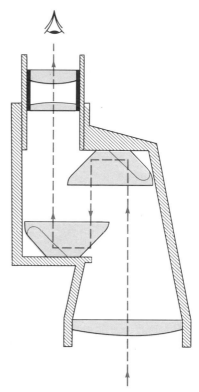

The simple telescopes used by Galileo used a convex lens to collect the light and a concave lens as an eyepiece. It gives an upright image but results in a very restricted field of view.

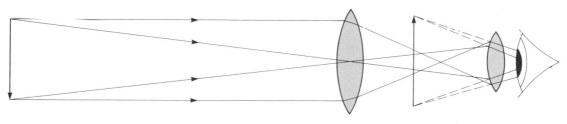

Kepler modified the design of Galileo's telescope by using a convex magnifying lens for the eyepiece. This improves the field of view but the image is then inverted.

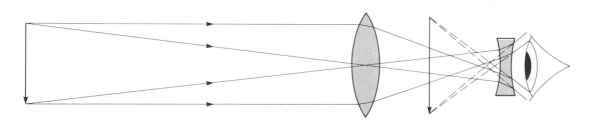

Hevelius' 150-foot telescope being erected at Gdansk, as illustrated in his *Machina Coelestis*, published in 1673.

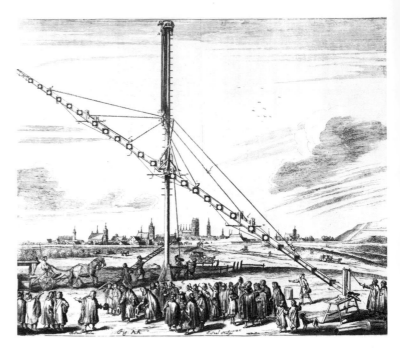

Paris, Cassini used long telescopes very successfully, observing markings on planets, the rotation of several planets, and the Cassini division in the rings of Saturn. But chromatic aberration was still a serious problem.

Newton believed that chromatic aberration could not be overcome in any lens system, and set out to design and build a telescope using a concave mirror instead of an objective lens. His famous reflector, demonstrated at the Royal Society in 1672, may have owed something to previous ideas by Mersenne, Descartes and Gregory, but it was undoubtedly the first successful example of the now classical Newtonian reflector telescopes. Success involves a more accurate optical surface on the mirror than is needed for a lens. It is very impressive that Newton, who is surely best known for his theoretical work on optics and on gravitation, should also have turned his hand to such an exacting task.

The reflecting telescope quickly replaced the refractor. John Hadley, for example, in 1721 made a reflector 15 cm in diameter and 1.52 m long, which was at least as good as a Huygens aerial telescope 20 cm in diameter and 37 m long. The main difficulty was

In a Newtonian reflecting telescope, the light is collected and brought to a focus by a concave mirror. The small flat mirror used to deflect the focus to an eyepiece outside the main tube does not interfere significantly with the light-gathering power or the quality of the image.

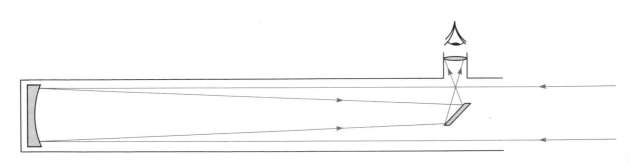

Newton's original reflecting telescope, which is now in the possession of the Royal Society in London.

the mirror, which was made of copper–tin alloys and some added zinc. This speculum metal could have up to 50% reflectivity, but soon tarnished.

The invention of the achromatic lens eventually reversed the trend towards reflector telescopes. It depends on the combination of two different types of glass which disperse the colours in different ways, cancelling out chromatic aberration, and it allows lenses with much shorter focal lengths to be used in telescopes and binoculars. It was invented by Chester Moor Hall, and was applied by John Dollond in 1758 and soon after by Clairault in Paris. The use of achromatic lenses allowed the length of refracting telescopes to be reduced dramatically: typically a telescope, formerly 10 m long, could now be built with a length of only 1 m.

Telescope mountings

The challenge of grinding and figuring a parabolic mirror for a telescope often carries an enthusiast into a project which is bigger than he appreciates. Apart from the optics, there is a rigid telescope tube to be made, and a smoothly acting bearing system on a rigid mount. The advantage over good binoculars is scarcely worth while unless the aperture is at least doubled, say to 75 mm for a refractor or 150 mm for a reflector, and it may be best to commit one's energies

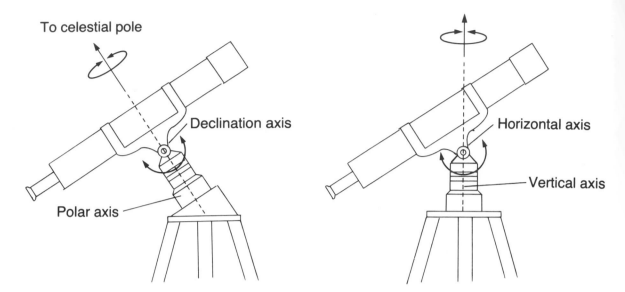

To be able to point a telescope anywhere in the sky, it has to be possible to turn it around two axes. The equatorial mount makes it possible to follow the motion of the stars across the sky by rotating the telescope around only one axis. An altazimuth mounting is simpler to construct and can now be used for very large telescopes because computers are used to position and drive them.

and funds to a more serious telescope such as a 250 mm reflector mounted on a permanent pillar and protected by a slide-off hut. You will then be able to participate in, for example, organised work on variable stars.

In deciding on the type of mounting you will be in good company. The problem of making the telescope move smoothly to compensate for the rotation of the Earth is ideally solved by using an equatorial mount, in which one axis is aligned with the Earth's axis. Incidentally this facilitates the use of hour angle and declination scales, which are very helpful in locating faint objects. For large modern telescopes the engineering problems dictate the use of the simpler altazimuth mount. Only since the advent of modern computers has this become feasible. Examples are the 6 m USSR telescope in the North Caucasus and the 4.2 m William Herschel telescope on La Palma. Herschel himself was forced to use the altazimuth mount for his big reflector telescopes, whose home-made speculum mirrors were up to 1.2 m in diameter. Amateur telescopes can be very effective with a simple but smoothly acting altazimuth mount. Observations near meridian crossing need very little elevation movement; the 1.8 m telescope constructed by Lord Rosse at Birr in Ireland could observe only within half an hour of meridian transit, a restriction which allowed the use of a mounting which was very simple in theory, but cumbersome in practice.

Amateur telescope builders and astronomers should not be deterred by any such difficulties, but should be encouraged by the remarkable successes, based on meagre resources, of the following amateurs of the past: Flamsteed, Herschel, Bessel, Lockyer, Hevelius, Moor Hall, Rosse, Huggins, Nasmyth, McMath, all of whose histories and accomplishments make inspiring reading.

Opposite The William Herschel 4.2-metre telescope in the Canary Islands. The positioning and movement of the telescope on its huge altazimuth mounting are controlled by computer.

Modern large telescopes

The mountains of the Canary Islands offer the best astronomical observing conditions in Europe, and the Roque de los Muchachos Observatory on La Palma, opened in 1985, has naturally attracted astronomers from several different countries to establish their various telescopes at this mountain-top site. In 1985 there were already six, and a seventh was under construction. All were different, and they by no means exhausted the range of possibilities. Why do astronomers need so many different kinds of telescopes?

The diversity is easily understood in the special cases of meridian astronomy and solar astronomy. In neither of these fields is it essential to use very large aperture telescopes, but other very special conditions dictate the form of the telescopes. Solar telescopes, for example, attempt to obtain photographs with the very finest detail; this can only be achieved if the telescope tube is evacuated, so a long vacuum tube is used. Meridian astronomy uses a transit circle, in which geometric precision is the vital consideration. But most of the telescopes are for stellar and extragalactic astronomy, where the first important distinction is between a wide-field imaging telescope, which is essentially a camera, and the so-called 'light collector', or 'light bucket', or on-axis telescope, whose purpose is to concentrate as much light as possible from a faint object into a point-like image which is then to be analysed by a

The international Roque de los Muchachos Observatory, La Palma in the Canary Islands. Mountain-top sites like this have the best observing conditions.

special instrument such as a spectrograph. The spectrograph analyses the wavelength of the light, displaying it as a spectrum. The reason for the division into on-axis and wide-field telescopes is, of course, the difficulty of building large telescopes with a large angular field of view.

The Schmidt camera is the classic example of the wide-angle telescope. Many of the photographs of star fields and nebulae in this book were taken with the 1.2 m Schmidt in Australia, and the classic survey of the northern sky was carried out by an almost identical Schmidt at Palomar Mountain in the USA. The primary mirror in a Newtonian reflector is paraboloidal, which gives only a narrow field; outside this field the images are poor, suffering badly from distortion. The Schmidt uses a uniformly curved ('spherical') primary mirror, which performs equally over a wide angle; correction for spherical aberration is achieved by a thin glass corrector plate mounted at the centre of curvature, i.e. at twice the focal distance from the primary. Whereas in a telescope using a parabolic mirror flat photographic plates are used at the focus, in the Schmidt telescope the focal surface is curved, and the photographic plates must be curved by clamping in a special holder. Schmidt telescopes cover a field of view of 7° or more.

Turning to the 'light collectors', whose design is essentially very simple, we find the main considerations are the practical engineer-

The distinctive shape of the McMath solar telescope at Kitt Peak in Arizona. The sloping tunnel through which the sunlight travels continues below ground.

The United Kingdom Schmidt telescope at Siding Spring in Australia is designed to take wide-angle photographs of the southern sky. The main mirror is 1.8 metres in diameter and the corrector lens is 1.2 metres across.

The Schmidt camera is designed purely for photography of wide areas of sky. It combines a spherical concave mirror with a thin lens of special design to produce an image that is accurately in focus over several degrees.

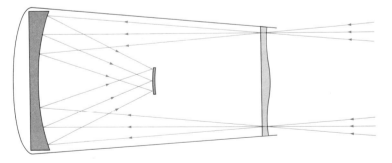

ing problems of mounting and guiding very large structures. The 3.9 m Anglo–Australian telescope, using a massive equatorial mount, can be set to a target object within about 10 arc sec, automatic guiding then holding to an accuracy of about 0.1 arc sec. The 4.2 m William Herschel Telescope, on La Palma, achieves a similar performance with the simpler and less expensive altazimuth mount. These telescopes are so much in demand that they are reserved for the most exacting observations, and telescopes like the 2.5 m Isaac Newton Telescope are used for most purposes.

The Anglo–Australian telescope, and its near-identical twin at Kitt Peak in Arizona, use primary mirrors that are not exactly paraboloidal. The secondary mirror, introduced by Cassegrain for directing the light back through a hole in the primary to a photographic plate or other detector, is usually hyperboloidal, but this is also slightly modified. The purpose is to increase the angular field over which good images can be obtained. The modification, proposed by Chretien and introduced by Ritchey (hence the designation 'Ritchey–Chretien') changes the primary from paraboloid to hyperboloid. Although a wide field, up to 40 arc min, can be obtained at the secondary (Cassegrain) focus, the prime focus cannot be used without a corrector lens.

Larger telescopes

Telescope designers have realised for many years that the heavy engineering of a modern reflector telescope is required only to support less than one gram of reflecting aluminium mirror surface, while the glass on which it is deposited weighs several tonnes and the telescope tube and mounting supporting the mirror may weigh

The 3.9-metre Anglo-Australian telescope at Siding Spring in Australia, in operation since 1974, was one of the first telescopes to be fully controlled by computer.

In a Cassegrain telescope, the main mirror has a central hole through which the light is directed after reflection from a secondary mirror.

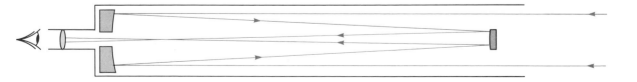

hundreds of tonnes. Could there not be some more economical means of collecting light, which would enable larger telescopes to be built?

Progress from the 5 m Palomar telescope to the USSR 6 m telescope was marked by difficulties in figuring the mirror and in maintaining its shape, particularly in changing temperature conditions. The advent of glass ceramic materials, such as Cervit, which do not expand at all when heated, has overcome this problem. Another solution has also been found: the mirror should be thin, so that its temperature quickly accommodates to the ambient air temperature. Using these ideas mirrors up to 8 or 10 m diameter may be possible, but the difficulty of transporting such a huge and delicate mirror to a mountain top, and of re-aluminising it later, has led to several radically different approaches.

A successful telescope using a cluster of separate mirrors, the Multiple Mirror Telescope (MMT) on Mount Hopkins in Arizona, can be used in two different ways. First, a series of mirrors can be used to combine the light from all six mirrors at a single focus. This is very demanding on the stability of the structure, but it is achieved by using an automatic optical alignment technique. Second, light

The Multiple Mirror Telescope on Mount Hopkins in Arizona uses six separate mirrors, each 1.8 metres in diameter, and concentrates the light they collect at a single focal point. The light-gathering power is equivalent to that of a single mirror 5 metres across.

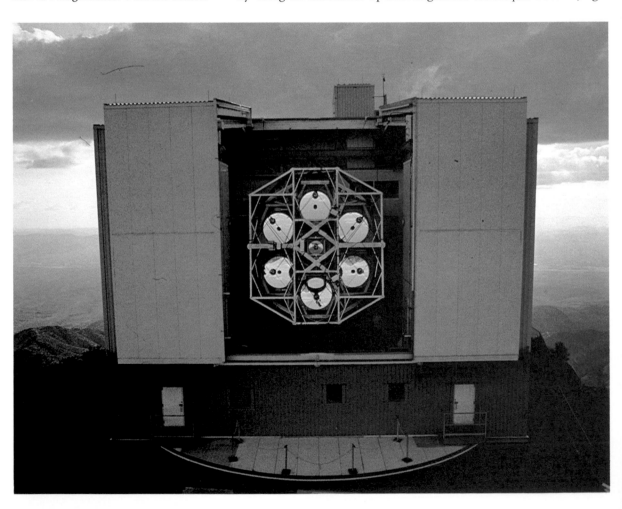

from the six focal points can be fed by an optical light fibre to a single detector, such as a photometer or a spectrograph. This type of telescope is economic to build, mainly because it can be very short, and needs only a small enclosure round it to protect it from wind and weather. Another approach is to construct a single large mirror from many segments, each individually figured as part of a paraboloid, and pieced together like a mosaic. The problems here are to figure the separate pieces as off-axis sections of a paraboloid, and then to set them to give a smooth surface, correct to about one-twentieth of the wavelength of light.

Finally, it has been proposed, particularly by Michael Disney, that it would be better to build an array of conventional telescopes, making up the light collecting area by sheer numbers. Each telescope would need its own spectrograph and other instruments, and the results would be combined after the observation.

Although the rewards for introducing these new techniques will be great, the difficulties and the cost will leave the present generation of 4, 5 and 6 m telescopes at the forefront of research for many years to come.

Telescopes in space

The advantages of siting an optical observatory on a mountain top instead of at sea level are immediately obvious. Often the observatory is above the clouds; even if there are no clouds there may be an atmospheric inversion layer trapping haze and pollution and leaving clear air above. Water vapour is much reduced, giving better atmospheric transmission, particularly for wavelengths just outside the optical spectrum. Thermal fluctuations, giving 'bad seeing' through random refraction, are much reduced. But all these effects are only reduced not eliminated: for that, it is necessary to operate a

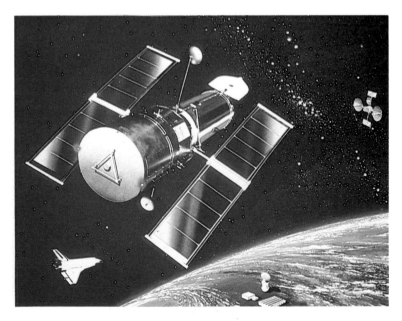

The Hubble Space Telescope is designed to give astronomers a clear view, unimpeded by the atmosphere. This artist's impression shows how it might appear in orbit.

telescope in space. Here the hopes of many astronomers are centred on the new Hubble Space Telescope, a 2.3 m diameter telescope orbiting 500 km above the ground. This telescope was to have been placed in orbit from a Shuttle spacecraft in the summer of 1986 but the disaster to the *Challenger* on 28 January 1986 caused a postponement until 1989.

The scale of the Hubble Space Telescope is hard to appreciate unless one has operated a comparable ground-based telescope, such as the 2.5 m Isaac Newton. But the complexity of the telescope and its instrumentation are even more impressive. A ground-based telescope has a team of engineers, instrument designers, computer experts and night assistants to run and maintain it. In space there is no equivalent of the day-time maintenance period, and instruments must be operated and changed by remote control. The rewards of success are, however, commensurate with the effort and cost. Star images sharp to better than 0.1 arc sec, over wavelengths extending far beyond the normally visible range, will open up a new window on the Universe. Unfortunately a solitary space telescope is not enough to satisfy a world-wide population of astronomers, and its use will necessarily be confined to selected very sensitive measurements, leaving the ground-based telescopes to carry most of the burden.

The electromagnetic spectrum

Most people can see colour in the light from several of the bright stars. Betelgeuse, for example, appears red and Rigel appears blue. (It happens that both authors of this book cannot see these colours in faint light: we are among the 10% of the male population who are partly colour-blind.) The colour of a star depends on how strong the starlight is in the different parts of the visible spectrum.

The range of colours present in sunlight was demonstrated by Newton, who placed a glass prism in a narrow shaft of sunlight.

An engraving in the Mansell collection represents Sir Isaac Newton investigating the properties of sunlight. He showed that a rainbow of colours (a continuous spectrum) is produced when sunlight passes through a glass prism.

This experiment is often repeated by accident when sunlight strikes table glassware or the bevelled edge of a mirror. It is not difficult to imagine that radiation from the Sun extends beyond this visible spectrum, beyond the red (the infrared) and beyond the blue (the ultraviolet). The infrared can easily be demonstrated in sunlight by testing the spectrum, spread out by a prism, with a thermometer with a blackened bulb. Most of the heat we receive from the Sun is in the infrared part of the spectrum.

Other familiar demonstrations of the spectrum of sunlight are the 'thin film' colours of soap bubbles and oil patches, and of course the colours of the rainbow. In astronomy the analysis of the spectrum of a star is carried out with a grating spectrograph. Some idea of the action of the diffraction grating in such a spectrograph can be obtained by using a gramophone record to reflect light from various sources at a glancing angle. Ignore the main reflection, and look for secondary, spread-out reflections. In this you can see the difference, for example, between continuous white light and the concentrated spectrum of a sodium street lamp. An even better demonstration can be had from an inexpensive pocket spectrograph, which looks like a tiny telescope but contains a small piece of diffraction grating.

The spectrum of starlight is of vital importance in astronomy. Colour, the balance of intensity across the spectrum, depends on temperature. The Sun's surface is at a temperature of 5800 K, giving a peak intensity in yellow light. Rigel's temperature is 11 000 K, giving a peak in blue light. The total amount of radiation from a star depends on its temperature and its surface area; if its temperature and distance are known, a measurement of its apparent brightness tells us the size of the star.

The spectra also contain detailed information on the concentrations of the different atomic elements that the star is made of. All elements have characteristic spectral lines, each with its own precise wavelength. The detection of a spectral line in light from a star immediately indicates the existence of a particular element in that star; the famous example is the discovery of helium in the Sun, through the observations of an unexpected yellow spectral line in sunlight.

Spectral lines can be seen either as bright emission lines against a dimmer continuous spectrum, which is usual for glowing clouds of gas such as the Orion Nebula and the nebulosity in the Pleiades, or as dark absorption lines in a bright continuous spectrum, which is usual in common stars such as the Sun. Absorption lines are seen when light from the hotter layers of the star has travelled through the cooler outer layers. The spectrum of sunlight contains many thousands of these dark lines, which are known as Fraunhofer lines after their discoverer. The spectra of bright stars such as Arcturus

Narrow, dark absorption lines cut across the continuous spectrum of the Sun at the particular wavelengths associated with the atoms in the Sun's outer layers.

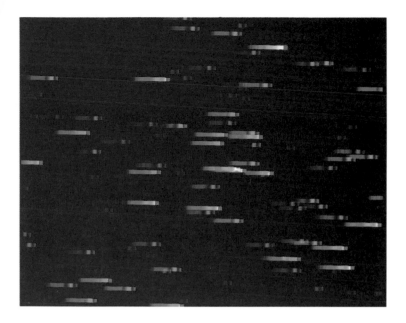

A large prism placed in the path of the light entering a telescope draws the image of each point of light out into a spectrum. This photograph shows the Hyades star cluster in the constellation Taurus. Careful examination reveals differences between the stellar spectra and enables their types to be determined.

have been mapped in great detail; the wavelengths of all the absorption lines agree with those in sunlight and in laboratory spectra, and we can confidently assume that the elements, and the laws of physics, are the same in these distant bodies as they are on Earth.

Atmospheric windows and radio telescopes

The atmosphere seems to us to be transparent, but it is in fact a completely opaque block to most electromagnetic radiation falling on the Earth. As a product of evolution, or design, or both, it happens that the range of wavelengths in the colours of visible light correspond to a transparent window, extending from about 350 nm to 1000 nm. (One nanometre, nm, is 10^{-9} m. The older designation 3500 to 10 000 Å, or angstrom units, is also used; 1 Å $= 10^{-10}$ m.) There is only one other completely transparent window, at much longer wavelengths. This is the radio part of the spectrum, extending from about 1 mm to 10 m. Between the optical and the radio windows lies the far infrared and the sub-millimetre region, whose boundaries are indistinct: some infrared astronomy can be achieved from mountain-top observatories at wavelengths as long as 20 or 30 μm (one micron, μm or μ, is 10^{-6} m), and some radio observations are possible at wavelengths near 350 μm (0.35 mm). The radio spectrum is filled with spectral lines, the vast majority of which are man-made. Tuning a radio receiver from one broadcasting station to another is radio spectroscopy; the amateur radio spectroscopist may wonder how any extra-terrestrial signals can ever be detected amongst such a concentration of artificial transmissions. It is as though dark-sky photography is to be attempted when the telescope is surrounded by a battery of arc lamps. Fortunately, by international agreement, certain bands of radio frequencies are kept free of transmissions so

Observations with the Swedish-ESO Submillimetre Telescope at La Silla in Chile will help fill the gap between the infrared and radio regions. This part of the spectrum is rich in signals from interstellar molecules. The 15-metre dish started regular operation in 1987.

that our radio telescopes can listen to the faint radio signals from outer space.

Many television enthusiasts already possess small radio telescopes in the form of parabolic reflectors, intended to receive television signals from satellites. Very sensitive receivers are necessary, as the signals are comparatively weak. If such a reflector antenna is directed towards the Sun, a large and noisy signal will be picked up, which can be heard as a hiss and seen as speckles on the screen. Radio telescopes collect radio waves from celestial objects, just as optical telescopes collect light waves. There is, however, a wider variety of radio telescopes, particularly because of the very wide range of wavelengths, extending over a range of 10 000 times or more.

The parabolic reflector, typified by the veteran 250 ft (76 m) Lovell Telescope at Jodrell Bank, collects radio waves over the whole of its area and focusses them on to a sensitive, low-noise radio receiver at the focus. The telescope can be directed to any part of the sky, and it can usually be seen at any time of the day or night following a fixed point in the sky, moving almost imperceptibly to compensate for the rotation of the Earth. The radio signals are analysed in various ways: the intensity at widely separated wavelengths gives the broad shape of the spectrum (like the colour of visible light); there may be a narrow spectral line, such as that from interstellar hydrogen gas which radiates at 21 cm wavelength

(a frequency of 1420 MHz), or there may be the regular heart-beat of a pulsar to follow and analyse.

Parabolic radio telescopes need accurate reflecting surfaces. As with optical telescopes, the accuracy depends on the wavelength; the surface must be smooth and the right shape to about one-twentieth of the shortest wavelength for which it is to be used. The Lovell Telescope will not work at wavelengths shorter than about 5 cm. The 100 m telescope at Bonn, and the 64 m Parkes telescope in Australia, work well at wavelengths below 1 cm. Millimetre wave telescopes, such as the 15 m diameter telescope on Mauna Kea, the high mountain in Hawaii, must have surfaces accurate to 50 μm. The largest reflector, 1000 ft (305 m) across, is at Arecibo

The Lovell Telescope at Jodrell Bank was completed in 1957 and remains one of the largest single-dish radio telescopes in the world.

The giant radio dish at Aricebo in Puerto Rico takes advantage of a natural depression in the Earth's surface. The dish itself cannot be moved but the antenna can be adjusted so that observations can be made over a limited band in the sky.

in Puerto Rico. This, like an optical Schmidt telescope, has a spherical rather than a paraboloidal reflecting surface. The reflector is fixed, so the radio beam can only be directed by moving the needle-like pickup antenna at its focus. Spherical aberration is compensated for by a phased array of dipoles along this pickup antenna.

At the longer wavelengths, radio telescopes are not necessarily reflectors. The familiar fishbone television antenna is, in principle, a radio telescope; an array of antennae of this kind, or even dipoles, can be constructed to cover several hectares (one hectare, ha, is 10^4 m^2). The discovery of pulsars, which depended on the use of a very large collecting area at a long wavelength, was made with a dipole array at Cambridge, England.

Resolving power and aperture synthesis

A good optical telescope on a good site can distinguish objects 1 arc sec apart; for example it could be used to read the centimetre scale marks on a ruler 2 km away. Some radio telescopes can improve on this angular resolution by a factor of 100 or even 1000. This remarkable feat is performed by the long baseline radio interferometer, which combines several large radio telescopes, a considerable distance apart, into a single receiving system. On the largest

The electromagnetic spectrum

scale the network of telescopes may span several countries stretching across the USA and Europe. On a smaller scale an array of telescopes may cover several kilometres; one such system in Cambridge, England, uses eight reflector antennae on a line 5 km long, while the most accurate is the Very Large Array (VLA) in New Mexico, which combines the output of 27 reflectors spread over three radial lines forming a Y shape.

The overall sizes of these interferometer systems are to be measured in terms of the wavelength at which they operate. If the wavelength is λ and the largest dimension is D, then the angular resolution is λ/D measured in radians (1 radian = 2×10^5 arc sec). Observations are often made, for example, at a wavelength of 6 cm. The VLA then has a resolution of 0.4 arc sec, and an intercontinental interferometer can have a resolution of a few milliarcseconds. All these systems can make maps of the radio emission from celestial objects such as radio galaxies and quasars: the longest baseline interferometers necessarily make rather crude maps, while the shorter baseline instruments make increasingly detailed maps as more radio telescopes are added to the system. One such system, covering baselines up to 200 km, is based on Jodrell Bank. It is the Multi Element Radio Linked Interferometer Network, or MERLIN for short: a fortunate acronym, as it has proved to have almost magical powers in drawing radio maps with an accuracy that was at one time thought quite impossible. These

The Very Large Array near Socorro in New Mexico is the world's most powerful radio telescope. Twenty-seven identical dishes, each 25 metres in diameter, are arranged along three arms in a Y-shape. Two of the arms are 21 kilometres long and the third 19 kilometres.

various interferometer systems perform as well, in terms of angular resolution, as an imaginary single radio telescope filling an area or aperture, whose diameter is the largest interferometer spacing. Map-making by this means is therefore often called aperture synthesis.

X-ray telescopes

Working towards shorter wavelengths than those of visible light, we first encounter the ultraviolet range, from about 350 to 100 nm. Telescopes for this range work on familiar optical principles, but they must be mounted on satellites and operated by remote control. In the far ultraviolet, from 100 down to 10 nm, and in the X-ray region, from 10 to 0.01 nm, there is the added difficulty that conventional mirrors and lenses cannot be used. Soft X-rays, with wavelengths near 10 nm, can be reflected from a highly polished metal surface at a grazing angle of incidence, and a pair of very elongated focussing surfaces can form a useful image. Alternatively a collimator system must be used: this consists of a set of metal grids arranged so that X-rays can only reach a detector from a very limited range of directions.

At this high energy part of the spectrum, the radiation more obviously arrives in discrete packets or photons. (The energy in a single photon is inversely proportional to the wavelength: a typical photon in the optical range has an energy of a few electron volts (eV), while X-ray photons have energies of tens or hundreds of kilo electron volts.) X-ray detectors record the arrival of individual photons. The simplest X-ray detector is a gas-filled tube in which an electric discharge is triggered by the arrival of a photon. The strength of the discharge can be made to be proportional to the energy of the photon, within an accuracy of about 20%. It is astonishing that the severe limitations on directional accuracy, due

The third Japanese X-ray astronomy satellite was launched in February 1987 and renamed Ginga, the Japanese for Galaxy, while in orbit. It is seen here before launch with solar panels extended.

to the difficulty of X-ray optics, and on energy resolution, have been overcome so well in satellite X-ray observatories such as UHURU, Einstein and EXOSAT. The technical difficulties are, however, balanced by the unexpected discovery of X-rays in many different types of celestial bodies, which we deal with in later chapters.

Infrared astronomy

On the long wavelength side of the optical spectrum, beyond the red light, lies the infrared range. Infrared wavelengths near 1 μm are used for 'seeing in the dark', when special cameras pick up the infrared radiation from warm objects. Longer wavelengths, beyond 20 to 30 μm, do not penetrate the atmosphere, so that infrared astronomy is divided between the 'near infrared', which can be used on ground-based telescopes, and the 'far infrared', which can only be used on space-borne telescopes.

The Infrared Astronomy Satellite (IRAS), launched in 1983, carried a telescope with a 60-cm mirror that was used to map the sky at infrared wavelengths. It is seen here in the space simulator at the Jet Propulsion Laboratory before launch.

The Infrared Astronomical Satellite (IRAS), showed how useful the far infrared spectrum can be. It operated only for one year, since its detectors were cooled with liquid helium of which there was a limited supply, but in that time IRAS surveyed the whole sky for infrared emitters of many different kinds, providing a fundamental catalogue of hundreds of thousands of objects.

Gamma-ray astronomy

The energies of gamma-ray photons, reaching 1 GeV (10^9 eV) and above, are such that a collision of a single photon with a thin sheet of lead will produce a shower of high energy electrons, all travelling in the direction of the original gamma-ray photon. This shower can be detected in a large gas discharge chamber, where it sets off a spark along the line of the shower. The direction of the spark can be found either by television cameras or by electronic means, using wire grids to locate the spark at various levels in the spark chamber.

In gamma-ray astronomy, which has been attempted with some success by means of apparatus carried in Earth satellites, the technical difficulties are compounded by a serious shortage of photons to be detected. The flux of energy from most sources is very low at these high energies and there are very few photons. A satellite gamma-ray telescope may therefore detect only a few photons per day. Surprisingly, a new technique becomes possible for energies exceeding a few hundred mega electron volts which has the twin advantages of operation on the ground and a very large collecting area. This is the atmospheric Cerenkov technique.

Cerenkov radiation is a flash of light generated by the passage of a high energy charged particle through the air. A high energy gamma-ray hitting the atmosphere generates a shower of electrons, which then emit a short but intense pulse of Cerenkov light. This pulse can be detected by photo-electric detectors having a wide field of view. Any high energy gamma-ray falling on the atmosphere within an area of 10 km^2 or more will be detected. This is the technique used to detect photons with 500 GeV energy from the Crab Pulsar.

5

The dynamics of the solar system

Let us forget for a moment the familiar picture of the planets, including Earth, in orbit round the Sun. Look instead at the sky, as it was seen by ancient civilisations, and watch the peculiar motions of the planets against the background of the stars. Mercury and Venus stay close to the Sun, sometimes leading and sometimes lagging in the eastward circuit. Mars, Jupiter and Saturn move more slowly to the east, taking 687 days, 12 years and 29 years, respectively, to complete a circuit. The eastward motion of all five is interrupted by periods of westward or 'retrograde' motion, at intervals of 116 days for Mercury, 584 days for Venus, 780 days for Mars, 399 days for Jupiter and 378 days for Saturn.

Today we know that these complexities cannot be reconciled with a solar system centred on the Earth and revolving around it, but until the 17th century the attempts to make a satisfactory fixed Earth model dominated astronomy. It was believed that heavenly bodies could only move with circular and uniform motion, and this led to ingenious solutions for the orbits of the planets. In these models the Earth was fixed at the centre of a rotating circle and points on this circle were the centres of smaller rotating circles representing the planets. Although by this means the phenomena of retrograde motion could be accounted for, many discrepancies remained between the observed behaviour of the Sun and planets and the predictions of the most complex epicyclic models.

Eventually, early in the 17th century, Galileo's first telescopic observations of the phases of Venus were made just at the time when Kepler was evolving the correct model of planetary motion. The combination of the varying crescent shape, and the variations of brightness of Venus, show very simply that Venus is in orbit round the Sun. Again, Galileo's observations of the satellites of Jupiter, obviously in orbit around that planet, gave a simple model for the whole solar system. It all seems so simple now: what then was the difficulty in discarding the Earth-centred model and adopting the 'obvious' heliocentric solution?

In the third century BC Heraclides of Pontus suggested that the motions of Venus and Mercury could be explained if they revolved around the Sun, and shortly afterwards Aristarchus of Samos formulated correctly the complete heliocentric idea. This idea was an affront to religious prejudices, and was contrary to the pervading Aristotelian cosmology. In particular it was strongly opposed by Hipparchus, the great and influential astronomer of that age who maintained that the circular motion of the Earth and planets

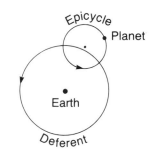

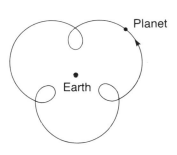

In an attempt to explain the observed paths of the planets in terms of purely circular motion, complicated schemes were devised in which the planets were envisaged to travel in circular orbits, called epicycles, the centres of which were simultaneously moving around the Earth in circular orbits called deferents.
Top: Epicycle and deferent.
Centre: Path supposedly followed by planet.
Bottom: Motion of planet seen in the sky from Earth.

Photographs of the planet Venus taken from the Lowell Observatory, Arizona, illustrate its phases and the range over which the planet's apparent diameter can change as viewed from the Earth. The variations in Venus's appearance arise from the relative motion of the Earth and Venus as they orbit around the Sun and because Venus' orbit is nearer to the Sun than the Earth's.

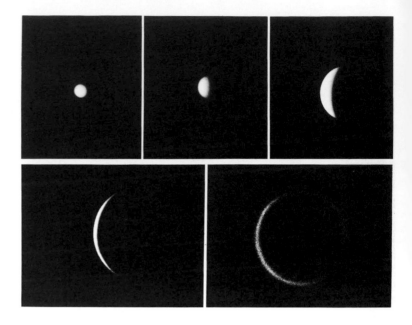

The phases and apparent change in size of Venus are readily explained if both planets orbit the Sun and Venus' path lies within the Earth's.

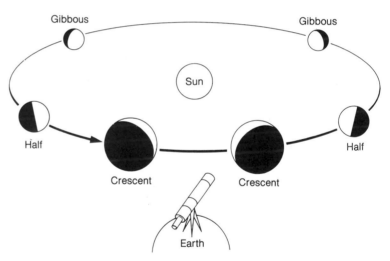

around the Sun could not account for the irregularities in planetary position. Indeed, in this he was correct, and when Copernicus revived the heliocentric theory in the 16th century he found it necessary to introduce many epicycles, and as far as the prediction of planetary positions was concerned his heliocentric solution with circular orbits was not greatly superior to the fixed Earth model of Ptolemy.

Kepler eventually reached the correct solution. He had available the accurate and systematic observations by Tycho Brahe, but only after long and painstaking attempts to preserve the simplicity and attraction of circular orbits did he conclude that the planets did not move in circular orbits. First, he realised that the planes of the planetary orbits pass through the Sun, and not the Earth. He was then able to formulate, in 1609, the following two laws:

Jupiter, with its four largest moons, Io, Europa, Ganymede and Callisto, photographed from the Lowell Observatory in 1915. These moons were first observed by Galileo when he turned the newly-invented telescope on the sky. They can be seen easily with the help of a small telescope or a pair of binoculars.

The illustration drawn by Kepler to explain his discovery that the orbit of Mars is elliptical and not circular, from his 1609 book *Astronomia Nova*.

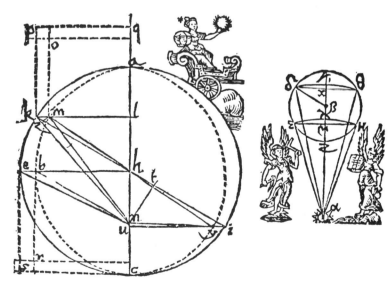

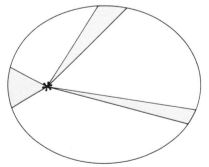

(i) the orbits are ellipses, with the Sun at one focus;

(ii) a radial line from the Sun to a planet sweeps out equal areas in equal periods of time.

Later, in 1619, Kepler added the third law:

(iii) the square of the period is proportional to the cube of the major axis of the ellipse.

Kepler determined purely from observations that the line joining the Sun to a planet travelling in an elliptical orbit sweeps out equal areas in equal times. This implies that the speed of the planet varies.

The physical basis to this geometrical pattern was provided 80 years later by Newton's law of gravitation. We now know that the inverse square law of gravitation is precisely followed over a range of distances from a few centimetres to the scale of the Universe. Strangely, it is a very weak force in comparison with electromagnetic forces; it predominates on large scales only because matter on the large scale is neutral so that electrostatic forces are unimportant.

The precision of the inverse square law can be illustrated by innumerable examples of orbits within the solar system. The return of Halley's comet after 76 years in a very eccentric orbit can be predicted within a few hours, and its position at its first reappearance within a few arc seconds; the only complication is to allow for the gravitational effects of the outer planets. Space probes can be sent in orbit round the Moon, or swung round the Moon or a planet to accelerate them to a new direction; the calculations use only the inverse square law of gravitation.

The description of the orbits of the planets is a very important step in the understanding of the origin of the solar system. All the orbits are traversed in the same direction (anticlockwise as seen from above the North Pole). The orbits are all approximately in the same plane; the main exceptions are the extreme inner and outer planets, Mercury and Pluto, whose orbital planes are inclined by 7° and 17°, respectively. These two are the smallest planets, and therefore most easily disturbed from an originally co-planar orbit. Most planets also rotate in the same direction; the exceptions are Uranus, whose rotation axis is almost in the plane of the ecliptic, and Venus, which is rotating very slowly in the retrograde direction. Like the inclined orbits, we regard these as exceptions due to a chance complication during the formation of the system as we now see it.

The satellites of the planets, starting with the Moon and increasing in number through Phobos and Deimos around Mars to the ten major satellites of Saturn, total 43. Like the Moon, all are rotating so that they always present the same face to their parental planet. Their revolution is generally in the same direction (prograde) as the whole system; notable exceptions are Phoebe (the outer satellite of Saturn) and Triton (a satellite of Neptune).

All these dynamical characteristics of the solar system indicate clearly an origin in a condensing gas cloud. The progression of condensation, through minute dust particles to progressively larger agglomerations, can be understood generally, but not in detail. What, for example, determined the remarkable progression in the radii of planetary orbits, which seem to follow a simple geometrical law? In 1772 Titius first recognised that a simple relationship existed, and the law was formulated mathematically by Bode in 1778. From this Titius–Bode law it appeared that the succession would be complete if there were another planet between Mars and Jupiter, at a radial distance of 2.8 times the radius of the Earth's orbit (known as the astronomical unit, or AU). In 1801 Piazzi, working at Palermo, discovered the minor planet Ceres which fitted into the gap, and today we know that there are about half a million much smaller bodies, the asteroids, moving in similar orbits. Is it natural to expect a complete succession of planets, and if so why did the 2.8 AU planet break up, or perhaps never form?

We will take another look at this question in Chapter 8, when we have collected together some further information on the composition of the planets.

Resonances

The locked rotation and revolution of the Moon, so that it always presents only one face to us, is a simple example of many resonances which occur between planets and their satellites. The rotation of Mercury, with a period of 89 days, is locked to its orbital period of 58.65 days: the ratio of periods is precisely 3:2. The mean orbital periods of Pluto and Neptune are in the same ratio. Io, Europa and Ganymede, three of the four brightest satellites of Jupiter, are locked in a three-body resonance. Even the orbits of the asteroids show the effects of resonances; their distribution in distance from the Sun shows gaps corresponding to periods in simple ratio with the orbital period of Jupiter.

Again, obviously, these are the effects of gravitational interaction. The details are less obvious, since the actual locking requires the dissipation of energy. Locking the Moon to the Earth is the result of dissipation of energy in tides on the Earth; probably the satellites of Jupiter followed a similar history. The larger scale effects of Mercury and Venus seem to require a further refinement of the overall condensation process.

Dynamical computation

Mathematicians know well that the orbits of two bodies in mutual gravitational attraction follow a very simple law, while the motions of a group of three defy analysis. How is it that any precision can be obtained in the predicted motion of the very many bodies in the solar system?

The simplicity of the whole dynamical system, and the success of Kepler's laws, depend entirely on the overwhelming effect of the Sun. Almost 99% of the mass of the whole system resides in the Sun; any small body, such as a space probe, is moving in the gravitational field of the Sun with only minor perturbations from the planets, except when it is very close to one of them. The planetary orbits are very good ellipses: they do in fact change slowly due to perturbations from the other planets, but there is no need for any complex multi-body analysis.

This comforting situation is again a pointer to the theory of the formation of the whole system. It is probable that complex gravitational interactions would result in orbits that would lead to disaster rather than stability. The structure of the solar system may be a good model on which to base our natural speculation about the existence of planets round other stars. Binary stars, for example, probably have no planets, and planets can generally be expected to be insignificant compared with their parent suns.

The discovery of Neptune

Herschel's discovery of Uranus was the result of an assiduous survey of the sky, without preconceptions of the nature of any

Urbain Jean-Joseph Leverrier (1811–1877), whose calculations led directly to the discovery of the planet Neptune in 1846, depicted on a commemorative medal.

unusual object that might be found. In fact, for some time he believed that he had discovered a new comet. The discovery of Neptune was, in complete contrast, the result of a prediction based on precise calculation. The orbit of Uranus departs appreciably from a pure ellipse, and the theory was that the gravitational attraction of an unknown planet was causing the disturbance. Calculations by John Adams in Cambridge and by Urbain Leverrier in Paris gave the correct prediction; Leverrier was the more successful since he advised the Berlin Observatory of his predictions, where in 1846 Johann Galle found the new planet at his first attempt.

This familiar story, which again illustrates the power of dynamical calculations based only on minor perturbations from the Sun's gravitational field, has recently acquired a new twist. A search through the notebooks of Galileo, who made an assiduous study of Jupiter and its four main satellites, shows that he was the first to observe Neptune, 233 years before its discovery by Galle. It happened that for about a month in 1613 Neptune and Jupiter were so close that they appeared together in the field of view of Galileo's telescope. He was a careful and precise observer, and he recorded the position of the slowly moving planet on several occasions.

Neptune has not yet completed one orbit since its official discovery, so it is very valuable to have good measurements available spanning more than two complete orbits. In fact, the positions measured by Galileo differ by about 1 arc min only from the positions based on modern observations. It is not yet clear whether this is the accuracy to be expected from the calculations, or whether some small perturbation from another planet is responsible.

General relativity

The precision with which Newton's laws of motion, and the inverse square law of gravitation, account for the dynamics of the solar system is made more remarkable by the discovery that these laws are only approximations. The general theory of relativity, formulated by Einstein by 1916, shows that gravity has a more profound effect than a simple inverse square law of force. It is difficult to detect the inadequacies of Newtonian mechanics in the solar system, but three important tests have shown unequivocally the correctness of general relativity. These demonstrate the effect of gravity on the propagation of light, and on the orbit of a planet in the varying gravitational field of an elliptical orbit.

The first measurement of the deflection of a light ray by the gravitational field of the Sun was made during a total eclipse of the Sun on 29 May 1919. The measurements were made by British expeditions to Sobral in West Brazil and to the island of Principe off the coast of West Africa. The measured deflection as determined from the position of a star seen close to the edge of the disc was 1.75 arc sec – in conformity with the prediction of the general theory. Other modern gravitational theories have been developed subsequently which predict slight differences in the value of the deflection compared with the prediction of general relativity. The most precise modern measurements using both optical and radio telescopes have all been decisively in favour of the Einstein theory.

The second effect on a ray passing close to the Sun is to increase the travel time of a light or radio pulse. Space probes emitting sharp radio pulses now travel in orbit round the Sun, and radar pulses are also sent in the double journey to planets at superior conjunction, i.e. on the other side of the Sun. In both cases delays of some hundreds of microseconds are observed, depending on the closeness of the ray to the surface of the Sun.

The other major prediction concerned an anomaly in the orbit of the planet Mercury. In the mid-19th century the French scientist Arago suspected that the orbit of Mercury was not in strict accordance with the predictions based on the Newtonian theory. He asked the young astronomer, Leverrier, to make a careful analysis of the orbit. This investigation revealed that the perihelion of the planet (that is the point in the orbit closest to the Sun) was changing year by year by 574 arc sec per century. The effects of perturbations by the other planets should lead to an advance of 532 arc sec per century and Leverrier concluded that there was an unexplained discrepancy of 42 arc sec per century in this precession. The effect is minute – nevertheless it was an outstanding discrepancy on Newtonian theory. Many reasons for the anomaly were suggested – for example, Leverrier predicted that the effect would be caused if a hitherto unseen planet existed in orbit around the Sun at a distance of 3×10^7 km – but no rational solution was obtained on the basis of classical mechanics.

Early in the 20th century the problem had worried Einstein, and soon after the publication of the special theory of relativity he believed that the geometry of curved space (Riemannian geometry) which he was exploring in the development of the general theory might account for the anomaly in the shift in the perihelion of

Mercury. When he eventually applied the equations of the general theory to the case of Mercury the prediction was that, over and above the motion due to planetary perturbation, the perihelion should advance by about 1/10 arc sec for each orbital revolution. Since the planet makes 420 revolutions per century the predicted relativistic advance of 42 arc sec per century was precisely the anomaly found by Leverrier.

The mutual perturbations of the planets in the solar system cause all the orbits to rotate in their own plane – the precession of the perihelion – and the relativistic effect is most prominent in the case of Mercury because of its proximity to the Sun. For the Earth the precession of the perihelion is 1165 arc sec per century (the perihelion point completes the orbit in 111 270 years). The relativistic effect accounts for only 3.8 arc sec per century. The agreement of the observed amount of precession of the perihelion with the predictions of general relativity for Mercury, Venus and Earth is an important observational verification of the general theory.

Are there more planets?

Although a multitude of problems concerning the solar system remain to be solved, the purely dynamical aspects appear to have reached a degree of finality, based on the Keplerian and Newtonian laws to a good degree of approximation, refined by general relativity in certain cases, for example in accounting for the anomalous precession of the inner planets. After the discovery of Uranus the irregularities in its motion led to the predictions by Adams and Leverrier that these must be caused by the perturbative effects of another planet. These predictions led directly to the discovery of Neptune. Later, more refined studies of the behaviour of Uranus led Lowell to predict that there must be another planet in orbit around the Sun beyond Neptune. This planet, Pluto, was discovered in 1930. The question is occasionally raised today that one or more further planets beyond Pluto must exist to explain certain discrepancies in the motion of the outer planets. In fact, no accepted discrepancies in motion significantly beyond the errors of measurement have been found. Similarly, cometary observations led occasionally to the suggestion that departures from predicted positions may have been caused by an undiscovered planet. Although it is not possible to be dogmatic about the possible presence of one or more undiscovered planets, two remarks may be made. No calculations based on supposed discrepanices of known planetary or cometary orbits have yet led to the discovery of a planet beyond Pluto, and the general astronomical opinion is that no further planets comparable to those already known remain to be discovered.

Opposite This crystal celestial sphere (known as the Powderham sphere, after its original home) was made in London in about 1748. The 15-inch sphere carries exquisite diamond engraving of the stars and constellation figures and, inside, there is a 5-inch Earth globe. The globes and engraved brass scales can be moved to demonstrate the apparent motion of the heavens.

A miniature orrery, in the collection of the Science Museum in London, made by Edward Troughton and dating from around 1800. Its diameter is 30 centimetres and it shows Mercury, Venus and the Earth.

Demonstration models

Although the solution of the major dynamical problem of the solar system has been largely outside the scope of amateur investigators, the motion of the planets has provided a rich field for model makers.

Globes to represent the stars in the sky were in use by ancient Greek astronomers. 'Skeleton' globes, made up of metal rings each representing an important circle on the celestial sphere, also have an ancient origin. They are known as armillary spheres. The Sun, Moon and planets were often represented as discs on movable rings and the armillary spheres were used for both teaching and observation. After the general acceptance of the heliocentric theory, particular models were constructed to illustrate the motion of the Earth and planets around the Sun. These models became known as orreries for the following reason. The first models were constructed in the early 17th century by George Graham, a notable instrument

maker of that time (and a nephew of Thomas Tompion the famous clockmaker). Graham's model was intended for Prince Eugene of Saxony, but before it was delivered the device was copied by John Rowley in about 1713. Rowley's patron was Charles Boyle, the fourth Earl of Orrery, but it was the essayist Sir Richard Steele, ignorant of Graham's priority, who announced that Rowley would call the machine an orrery to honour his patron. The machine could be operated by turning a handle – one turn represented a day and the model was essentially limited to illustrating the motion of the Earth–Moon system around the Sun. The orreries soon became more complex and complete. Although it was impossible to simulate the scale of the solar system, the planetary motion could be reproduced satisfactorily. The planets were carried on radial arms and made to rotate around the central Sun by a system of gearing at the correct rate. Later models also included the known satellites of the planets in appropriate motion around the planet. In some forms the models were called planetaria, but today that word is generally used for the special dome-shaped building and projector that demonstrate the motion of the stars and planets as seen by an observer on Earth.

The extent of the solar system

Light from the Sun takes eight minutes to reach the Earth 1 AU away; light from the nearest star takes over four years. At what distance, measured in AU, does the solar system end and interstellar space begin? Despite the fact that the gravitational pull of the Sun goes on, in a sense, for ever, there is a real meaning to the question. Let us rephrase it: what is the maximum radius of an orbit in which a body is still gravitationally bound to the Sun?

Computation shows that the stars of the Milky Way, taken as a whole, would disrupt orbits more than 80 000 AU in radius. Strong individual stars would occasionally disrupt orbits somewhat closer, say at 50 000 AU. This must be regarded as the limit of the solar system; it is at this large distance, about as far as the nearest star, where we believe the most distant members of the system may be located. These are the comets, lurking in outer space, totally unobservable until one is disturbed, possibly by a passing star, into an orbit which sends it almost straight towards the Sun.

This cloud of potential comets was proposed by J.H. Oort; it is therefore known as the Oort Cloud. We know little of it, but if it exists it was probably much more populated in the early years of the solar system. Disruption would then be more probable; it may be that in an earlier era a veritable bombardment of comets reached the inner solar system, leaving its mark on such unprotected surfaces as the pock-marked face of the Moon.

6

The Moon

Galileo, the first person to see the Moon through a telescope, was a fortunate man. The unaided eye cannot tell us that the light and dark features of the Moon's surface are mountains and plains: Galileo called them *terrae* and *maria*, thinking that the plains must be seas. Even through modest binoculars we can see bright peaks on the inner edge of the crescent moon, the shadow edge called the 'terminator' (that is the boundary between the lighted and dark regions of the Moon). A 20 cm aperture telescope reveals a wealth of detail. The entire surface of the Moon, including the maria, is riddled with craters. It is surely one of the most fascinating celestial objects to study through the telescope.

Despite its importance in generating the Earth's tides and in

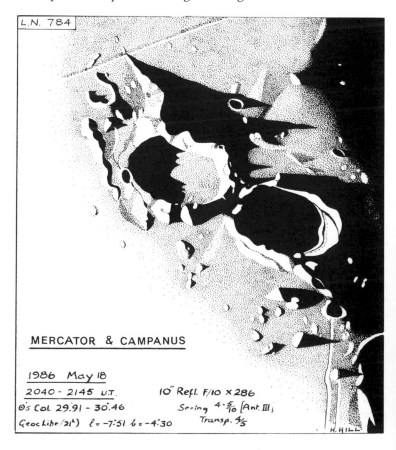

This striking drawing of the lunar craters Mercator and Campanus was made by the British amateur observer, Harold Hill, who used a 10-inch reflector.

night-time illumination, the Moon is a minor astronomical body, with a mass just over 1% that of the Earth. At a mean distance of 384 401 km, and with a diameter of 3476 km, it subtends an angle of just over half a degree: by coincidence the same angular width as the Sun. If the Moon were as far away as the Sun it would be only 5 arc sec across, and little detail would be seen even in a large modern telescope. As it happens, it is near enough and bright enough for its surface features, and its complex motion, to be studied in great detail. It is also near enough to be visited by man.

Until close-up photography of the Moon became possible from space vehicles in the 1960s, the most comprehensive map of the Moon, showing over 50 000 craters, had been made by an amateur. It is still a very rewarding exercise to sketch features of the Moon as seen through the telescope and to try one's skill as a photographer. (Any amateur society would help with details; as a start try a fast film at 1/100 s, f/8, and then progress to longer focal lengths.)

Lunar atlases are readily available from which large numbers of the maria, mountain regions and craters that have been charted and named can be readily identified. An easy guide to some of the major lunar features may be given by referring to the position of the terminator.

If the new crescent Moon is inspected through binoculars the outline of the whole disc may be dimly seen. This is made visible by the sunlight reflected from the Earth – the earthshine. A few days after new moon the terminator crosses one of the extensive dark areas – *Mare Crisium* extending for several hundred kilometres and encompassing some 106 000 km^2 of the lunar surface. At this stage some of the larger craters become visible; those south of the *Mare Crisium* are more than 125 km in diameter and a few thousand metres deep. About five days after the new moon when *Mare Crisium* is wholly visible a chain of some of the Moon's greatest craters are illuminated. These are near the dark area known as the *Mare Nectaris* and are heavily shadowed by the immense walls around their rims. For example, the crater *Theophilus*, with a diameter of over 95 km has terraced walls rising 5.5 km above the interior of the crater. A day later the terminator reaches another of the major dark areas – the *Mare Tranquillitatis*.

At the first quarter, or half moon, the terminator reaches one of the most spectacular features of the lunar surface, the range of mountains known as the Apennines with peaks of 7 km above the surface bordering the major dark area of the *Mare Imbrium*. The area of this sea equals that of France and the UK together. At this stage a chain of great craters becomes visible, one of which, *Ptolemaeus*, has a diameter of 160 km. To the south, binoculars will reveal one of the largest of the lunar craters, *Clavius*, with a diameter of 240 km. The terminator soon reaches the great crater of *Copernicus* visible between the *Mare Imbrium* and the large area of the *Mare Nubium* to the south. Three days after half moon the terminator reaches the largest of the lunar seas – the *Oceanus Procellarum*. A very bright object can be seen in the *Oceanus Procellarum*, the crater *Aristarchus* 50 km in diameter and 1.5 km deep. Shortly before full moon the terminator reaches the Moon's greatest crater, *Schickard*, 216 km across with irregular walls peaking to 2.7 km above the floor of the crater.

The Moon, 7 days after new Moon (first quarter) and on the wane after 20 days. The sharp contrast between the dark mare areas and the lighter, heavily cratered regions is very marked. Along the terminator, where it is dawn or nightfall on the Moon, dramatic long shadows cast the features into sharp relief.

At full moon the absence of shadows makes identification less easy but systems of bright 'rays' which emerge from the crater regions and may extend for hundreds of kilometres become prominent. These rays are believed to be reflected light from finely pulverised material thrown out from the crater. Those from the crater *Tycho* which is near the crater *Clavius* referred to above can easily be seen through binoculars.

The orbit of the Moon – libration

The period of the Moon's rotation, 27.3 days, is the same as the period of its revolution in orbit round the Earth. We therefore see the same face of the Moon all the time, and until the advent of lunar orbiters we had little idea of the nature of the surface of the other half. There is, however, a slight apparent wobble, known as libration, which brings a total of 59% of the lunar surface into view. Libration in longitude is due to the ellipticity of the Moon's orbit, which allows us to see it sometimes a little ahead, and sometimes a little behind, its mean position. Libration in latitude is a result of a misalignment of the Moon's rotation axis with respect to its orbit, by 6°41'. This periodically allows areas of the lunar surface beyond the north and south poles to be seen from Earth. There are also additional minor librations which arise from non-sphericity and the slightly asymmetric mass distribution in the Moon.

Libration brings about a continual change in the profile of the Moon's limb, as different mountains and craters come into view. This produces interesting variations in occultations of stars, particularly those which just graze the limb of the Moon. Very fine detail of the Moon's motion can be obtained from grazing occultations; they can, however, only be observed from locations defined within a few kilometres, and a precise prediction must be used.

Early in the 17th century Edmond Halley discovered that every 18.61 years the Moon occulted stars in similar regions of the sky. The explanation is that perturbations of the lunar orbit lead to a revolution of the line of intersection of the plane of the Moon's orbit with the ecliptic with this period of 18.61 years. This is known as the nutation period, that is the period of the Moon's node. Other recurrent features that have been identified and which arise from the complex influences on the Moon's orbit are the Metonic and Saros cycles. The Metonic cycle of 19 years is the period when a full moon occurs on the same date – a significant parameter in the calculation of the date of Easter. After a period of 18 years 11.3 days the Earth, Sun and Moon return to almost the same relative positions. Thus an eclipse is likely to be followed by another eclipse after 18 years 11.3 days. This interval is known as the Saros period or cycle and was used by ancient astronomers to predict eclipses with a fair degree of accuracy.

The exploration of the Moon by manned and unmanned spacecraft

Although accurate maps of the lunar surface have been produced by telescopic observations from Earth, the limit of resolution at the

Apollo 17 astronaut, Harrison H. Schmitt, collects samples at the Taurus-Littrow landing site on the 11th December 1972. The rake he holds was used to collect samples of rock ranging in size from 1.3 to 2.5 centimetres across.

Moon's distance even with a large telescope is about 200 m. Many controversies arose about the real nature and origin of the surface features. For example, had the craters been caused by volcanic eruption or by meteoric impact, and was the surface hard rock or covered with thick layers of dust? How old were the lunar rocks? Was there any appreciable magnetic field on the Moon?

Many of these queries have been answered by the series of manned and unmanned spacecraft sent to the Moon or placed in orbit around it, since the Russians first succeeded in hitting the Moon with the *Luna 2* rocket on 13 September 1959, and transmitting to Earth the first photographs of the hidden side when *Luna 3* made its circumlunar flight. The first close up photographs of the lunar surface were provided by the American *Ranger 7* spacecraft, which continued to transmit photographs to Earth until the instant of its crash landing. This successful flight was made in August 1964, and a selection of the close up photographs was rushed to the General Assembly of the International Astronomical Union, then meeting in Hamburg. At the moment of their arrival a large meeting of lunar experts was in progress and the instant reaction of the various authorities on the Moon epitomises the vagueness of our knowledge at that time. One of the present authors was at the meeting and remembers the extraordinary atmosphere created when, instantly, Professor Gold of Cornell University and Professor Kuiper of Tucson, Arizona, rose to give their opinions on the photographs. Gold maintained that the photographs confirmed his view that the Moon's surface was covered with so much dust that an astronaut would sink at least up to his knees on landing, whereas Kuiper said that the photographs made it clear that the surface was of hard solid rock with no significant layer of dust.

The extremes of this argument were settled when the Soviets succeeded in making an automatic soft landing of *Luna 9* in February 1966. Clearly this probe did not sink into deep dust and it transmitted to Earth detailed photographs of the terrain in its vicinity. The landscape looked very much like a lava plain with rocks and craters everywhere. A series of American lunar orbiters succeeded in making a systematic photographic survey of the whole surface with a resolution of 2 m by August 1966. This survey and the soft landing of the American *Surveyor* probes prepared the way for the epic manned flights to the Moon. Beginning with *Apollo 11* in July 1969 there were six manned landings on the Moon, and 382 kg of lunar samples were returned to Earth for detailed analysis. Simultaneously the Soviets returned samples automatically and also made a detailed inspection of considerable areas of the lunar surface by means of their roving *Lunakhod* vehicles under control from the Soviet Union. Further lunar orbiters have effectively completed the detailed photographic survey of the lunar surface, and the instruments, which remained on the surface after the manned expeditions, transmitted seismographic and other important scientific data to Earth for several years.

The geology of the Moon

The basic materials of the Moon and of the Earth are similar, but their surfaces are very different. On Earth the existence of a protective atmosphere, the constant action of water and the steady drift of the continental masses ensure that sea and sedimentary rocks cover most of the surface. The Moon's surface is static and unprotected. Only a constant rain of meteorites can change the lunar landscape. The new data from the manned and unmanned spacecraft can therefore reveal the history of the Moon from the time of its solidification. There are no young rocks; the highland rocks are 4 billion (4×10^9) years old, and the maria rocks are 3.2 to 3.8 billion years old. Significantly, both these ages are almost, but not quite, as great as those of the oldest rocks on Earth.

The maria are less cratered than the highlands; they are younger and did not experience an earlier era of bombardment by meteorites. Analysis of material from the maria showed that they were formed from molten rock which filled low lying areas with lava. The material is similar to terrestrial basalts, and is undoubtedly of volcanic origin.

Samples of rock from the highlands show that this original lunar crust is also basalt, very similar to the rocks of the maria. The difference in colour comes from ilmenite, a dark form of basalt which is much more abundant in the maria.

The discovery that both the highlands and the maria are covered in basalt implies that at some stage in history the whole surface of the Moon must have been molten. Originally this must have occurred in the final stages of accretion, when material was still falling on the surface of the newly formed Moon. After this bombardment, the surface cooled and solidified, leaving molten rock beneath. The molten layer extended some hundreds of kilometres below the surface, and was appreciably enhanced by the heat generated in the radioactive decay of heavy elements. This deeper molten rock then emerged as lava to fill the basins of the maria.

The lunar farside, which is never visible to observers on the Earth, is relatively devoid of dark maria compared with the face we can see but the crater Tsiolkovsky, flooded long ago with molten rock, stands out starkly black in this Apollo orbiter image.

The final solidification of the Moon occurred about 3 billion years ago. Since that time the surface features have been modified only by the bombardment of meteorites, which fall unimpeded by any atmosphere. Big meteorites are rare, but the constant rain of micrometeorites will eventually wipe out small surface features, such as the shallow footprints left in the dust by the Apollo astronauts.

The belief that the Moon has been internally quiescent for a long period is supported by the data transmitted to Earth from the seismometers left on the Moon by the *Apollo* astronauts. These data have shown only very minor seismic activity. The recorded moonquakes release energy less than one-billionth of that involved in earthquakes. Further, the data indicate that these minor disturbances occur at least 800 km below the surface of the Moon. The conclusion is that if the Moon now has a molten core it extends only a few hundred kilometres from the centre and that the outer regions consist of a thick shell of rigid rock at least 800 km deep.

Artificial tremors of the Moon have been stimulated by the impact of the third stage of a *Saturn V* rocket and by the landing modules of the *Apollo* flights. The analysis of the seismometer records has given an indication of the speed of transmission of seismic waves through the Moon's interior. One interesting conclusion is that the crust of the Moon is thinner on the hemisphere facing the Earth than on the hidden face. The photographs from the orbiting spacecraft showed that the maria on the hidden side of the Moon

The main feature in this Apollo 17 orbiter photograph of the lunar farside is the crater Van de Graff.

are rare compared with the 35% of the surface that they occupy on the side facing Earth. The explanation may be that the thicker crust on the hidden face has inhibited the volcanism and lava flow that formed the maria on the visible face of the Moon. The differences in the thickness of the crust on the two hemispheres of the Moon is presumed to be related to the synchronism of the rotation and orbital revolution of the Moon which causes the same face always to be turned towards the Earth.

Another interesting geological feature of the Moon has been discovered from precise measurements of the spacecraft in orbit around the Moon. Whereas the theoretical acceleration due to the Moon's gravity is 162.7 cm s^{-2} (one-sixth of terrestrial gravity) the measurements showed that there were surprising localised variations amounting in some regions to about 0.2 cm s^{-2}. The variations are nearly all positive, and the conclusion is that they are associated with concentrations of mass, called mascons, coinciding with the basalt filled basins of the Moon.

The spacecraft measurements confirmed that the Moon has no general magnetic field, although a fossil magnetism has been found in certain rocks. The origin of this magnetic field is obscure; it may, like the Earth's magnetic field, have originated in the molten rock of the interior and faded away as the Moon solidified.

The origin of the Moon

Although the manned American and unmanned Soviet excursions to the Moon have given a broad picture of how radioactive heating, volcanic activity and meteorite bombardment may have formed the surface of the Moon as we see it today, the question of its origin remains obscure and debatable. However, the dating of the rocks returned to Earth by the astronauts gave one decisive result in this connection. The oldest rocks from the highland regions are 4.2 billion years old, and small fragments 4.6 billion years old were found. This leaves little doubt that the Moon and the Earth were formed at about the same epoch in the embryonic solar system but leaves open several theories of the process by which the Moon became a satellite of the Earth.

A group of theories which can now be rejected suggested that the material of the Moon was once part of the Earth. George Darwin (son of Charles Darwin) proposed that the Moon was wrenched from the Earth by combined centrifugal force and the gravitational attraction of the Sun. A later proposal was that the Moon was spun off the equator of the Earth not as a single piece but as a circular ring of gaseous matter that subsequently condensed into the solid body of the Moon.

The modern theory of the origin of the planetary system (Chapter 8) – that the whole solar system condensed out of a flattened rotating nebula – offers a better solution. Here the Moon would at one stage be a proto-planet, either orbiting the Sun on its own or as a part of a local condensation including pieces of the Earth and perhaps also of Mars. A random process of collisions during the final coalescence would then determine the fate of the Moon, following a similar pattern to the condensations of the satellites of the other planets.

7

The Earth

In the Space Age it seems absurd to suggest that the Earth is flat, even though it appears so when we look at the horizon from sea level. The higher we go, the easier it is to see the curvature of the horizon; it is clearly seen from a high-flying aircraft, and it is obvious to astronauts standing on the Moon. Nevertheless, let us imagine ourselves to be living among the Ancient Greeks, and let us try to prove that the Earth is a sphere.

First we follow Aristotle, who remarked that during a lunar eclipse the shadow of the Earth on the Moon is an arc of a circle, and not a straight line. We might then try to measure the curvature directly, for example by sighting along a straight canal; the surface should depart from a straight line by 9 cm in a distance of 1 km. This does not work very well, as refraction spoils the line of sight. We therefore follow another Greek, Eratosthenes, who compared the direction of the vertical at Alexandria and Syene, two places in Egypt 800 km apart. At Syene (close to the modern Aswan) the Sun is exactly overhead at midday in mid-summer, while on the same day at Alexandria it is 7.5° to the south. Eratosthenes obtained a value for the circumference of the Earth in stades, believed to be a unit of 157.5 m; if the conversion is correct, his value is 39 400 km, within 2% of the modern value of 40 074 km (24 900 miles) for the equatorial circumference.

Is the Earth a perfect sphere? If it were so, the distance between lines of latitude equally spaced in angle would be the same at all latitudes. In fact the distance is greater at the equator than at the poles. This shows that the Earth is flattened at the poles and bulging at the equator; it is an oblate spheroid, with polar diameter 12 713 km and equatorial diameter 43 km greater.

With the launching of Earth satellites in 1957 an entirely new technique became available for measurements of the shape of the Earth. From accurate tracking measurements of the satellites it became clear that the orbits were influenced not only by air drag but also by the non-spherical components of the Earth's gravitational field. The equatorial bulge exerts a torque that causes the satellite orbit to precess about the polar axis. It was soon discovered that the difference between the polar and equatorial diameters was 150 m less than had previously been thought. More surprising was the discovery that the Earth is not precisely symmetrical along its rotation axis. A depression of 30 m at the South Pole and a hump of 10 m at the North Pole led to the description 'pear shaped', even though the shape of a real pear is normally very much further from a true sphere. Several other irregularities have been revealed, for example a depression of 113 m south of India and a hump of 60 m near Great Britain. By means of many thousands of photographic and laser measurements of satellite positions and complex computer solutions in the 1960s the Smithsonian Astrophysical Observatory defined a Standard Earth which is believed to be accurate to 10 m, representing a 15-fold improvement in accuracy compared with the precision obtained before satellites.

The rotation of the Earth

We go about our daily business as though the Earth is at rest. We refer to the motion of the Sun and stars across the sky as the ancients did, although we have known for nearly 400 years that this diurnal motion of the heavenly bodies is an apparent movement arising from the rotation of the Earth. The fact that the Earth is not a perfect sphere but is flattened at the poles is a direct consequence of this rotation. Every particle is subject to the gravitational attraction of the rest of the Earth. If the material were distributed uniformly through a body of the size of the Earth at rest, it would be a perfect sphere. But, in the case of the Earth, because of the daily rotation there is a centrifugal force acting perpendicularly to the axis of rotation. This is zero at the poles and greatest at the equator, where it acts in opposition to the gravitational force. If the material of the Earth behaved as a uniform fluid then the result of this combination of the gravitational and centrifugal forces would be to flatten the sphere into an oblate spheroid. The potential energy of a body is the same at all points on this ideal surface; there is therefore no tendency for the oceans to flow from poles to equator or vice versa. However, the force of gravity does vary over the surface; the length of a seconds pendulum is 0.4% shorter at the equator than at the poles, corresponding to the smaller effective gravity resulting from the combined effect of the increased radius and the centrifugal force of the Earth's rotation.

A remarkable demonstration of the Earth's rotation was first made by Foucault in 1851. He suspended a pendulum, made from a heavy bob and a long wire, from a free pivot at the top of the dome of the Pantheon in Paris. When this pendulum was left to swing in a straight line, its plane of swing slowly rotated, or, as Foucault pointed out, the pendulum stayed in a fixed plane while the Earth rotated underneath it. A Foucault pendulum at the North or South Pole would make a complete rotation with respect to the Earth in 23 h 56 m, which is the rotation time of the Earth with respect to the stars. At the Equator there would be no rotation, as the point of suspension of the pendulum would be perpendicular to the Earth's rotation axis; at intermediate latitudes the rotation rate varies as the sine of the latitude. In London the period of rotation is 31 h.

This simple experiment is well worth a trial, but it does need a heavy pendulum bob, a symmetrical support, and at least 10 m height. Impressive examples can be seen in many museums and exhibitions, notably in the entrance lobby of the United Nations building in New York.

In Chapter 2 we mentioned the phenomenon of the precession of the Earth's axis of rotation. As first noticed by Hipparchos in

The crescent Earth hangs dramatically over the lunar horizon as seen from an Apollo orbiter.

120 BC, the position of the celestial pole with respect to the stars is not constant. Every 26 000 years the axis of the Earth's rotation effectively sweeps out a cone with an angle of 23°27'. In a few thousand years time the pole star will not be Polaris but will be in Cepheus. The precession is a consequence of the oblateness of the Earth. The equator of the Earth is inclined to the ecliptic plane, and because of the equatorial bulge the direction of the gravitational attraction of the Sun does not pass through the Earth's centre of mass. Thus the Sun's gravitational force attempts to pull the equator of the Earth into the ecliptic plane. However, the Earth is rotating and the couple generated by this angular momentum and the gravitational attraction of the Sun results in the precession of the axis of rotation. There is a similar, although much weaker, effect arising from the gravitational pull of the Moon on the equatorial bulge. In accurate astronomical measurements the gravitational effect of the

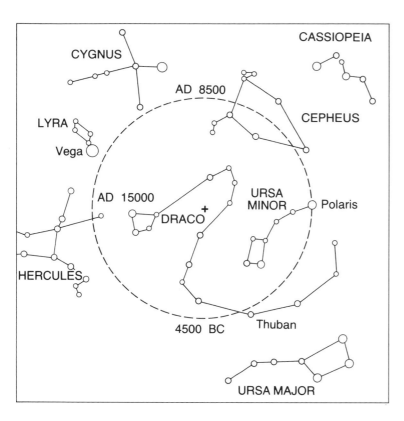

The Earth's rotation axis is not fixed in space. Its direction relative to the stars changes so that it sweeps out a circle in space every 26 000 years. This means that the north celestial pole, presently near the star Polaris, appears to trace out a circle in the sky in the same period of time.

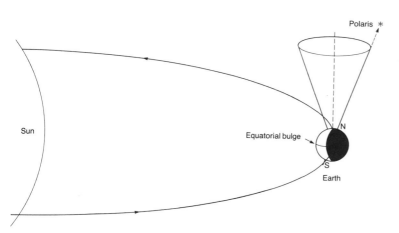

The precession of the Earth's rotation axis arises from the gravitational pull of the Sun (and to a lesser extent the Moon) on the Earth's equatorial 'bulge'.

planets on the equatorial bulge has also to be taken into account, although this is a minor contribution to the precessional effect compared with that of the Sun and the Moon. Precise absolute astronomical measurements have to take account of other minor irregularities on the precession. These are known as forced nutations and they arise because the orbits of the Earth and the Moon are slightly eccentric and because the plane of the Moon's orbit relative to the ecliptic varies. Thus, there is a periodic variation of the solar and lunar couple. The main effect is an amplitude variation of 9 arc sec with a period of between 18 and 19 years.

There is a further small effect in addition to the main precession and nutation arising from the existence of the equatorial bulge. The Earth's axis of rotation moves around the axis of inertia and this gives rise to an apparent wobble of the celestial pole with a period of about 1 year. From Earth the effect is that the trajectory of the pole is a minor spiral around the main circular motion. The amplitude is very small, about 1/10 arc sec, and is significant only in absolute measurements in astrometry.

The length of the day

If we observe the time at which a star crosses our meridian night by night we find that the interval is 23 h 56 m. If we make a similar observation on the Sun the average interval is 24 h. The first interval is known as a sidereal day and the second, by which we live our daily lives, is the solar day. The difference arises because the Earth rotates on its axis once in 24 hours but orbits around the Sun in the same direction as its rotational motion. Thus the solar day is longer than the sidereal day, and in one year there is one more sidereal day than there are solar days. In observatories it is sidereal time which effectively determines the pattern of observations and rate of movement of astronomical instruments; the sidereal and solar clocks coincide in time only once per year.

Although for practical convenience we divide every solar day into 24 hours the actual length of the solar day varies throughout the year because of the ellipticity of the Earth's orbit around the Sun and the inclination of the Earth's axis of rotation to the ecliptic. The mean solar time which we use for practical convenience is obtained by making all the days equal to the average length of a solar day. The timing of a day based on the real Sun is known as the apparent solar time, and the difference between this and the mean time which we use is the equation of time. The difference between apparent and mean time can be considerable, reaching extremes on 1 November of + 16 minutes (apparent minus mean time) and of − 14 minutes in mid-February. A conventional sundial will show apparent solar time, and if it is required to show mean time it has to be marked and calibrated using the equation of time.

The Earth's rotation is no longer used in accurate measurements of time. Atomic clocks, using the resonance of caesium atoms, provide a more uniform time scale which is adopted internationally. Against this atomic time scale the Earth's rotation is seen to be slowing down; this is due to ocean tides, which dissipate the rotational energy. The tides are due to the gravitational fields of the Sun and the Moon, but mainly of the Moon. The gravitational force of the Moon diminishes across the Earth; the side nearest to the Moon experiences the largest force, pulling the ocean towards the Moon. On the opposite side the force is the lowest, effectively allowing the ocean to move away from the Moon; the result is the familiar pattern of two high tides per day. The consequent slowing down increases the length of the day by about 1 ms per century. It happens that the universal, or atomic, time scale is appreciably different from the scale based on mean solar time, and that the cumulative effect of the slowing of the Earth's rotation is increasing the difference. Universal time is therefore corrected at fairly regular inter-

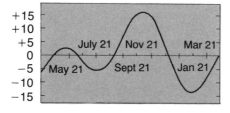

Minutes

Because the Earth's orbit is not circular, the apparent speed with which the Sun travels through the sky against the background of stars is not uniform. The result is that time measured with a sundial can differ from 'mean' time by several minutes. The difference between the two, the 'equation of time', varies through the year.

vals to bring it back into line with Earth rotation; it is now usual for a 'leap second' to be added to the time signal at the beginning of the year, and the same correction is often necessary in the middle of the year.

Another effect of the tides is a reaction back to the Moon, whose orbit increases slowly in size as the Earth's rotation slows down. In this way angular momentum is transferred from the Earth itself to the Earth–Moon system, and the total angular momentum is conserved.

In addition to this steady slowing there are irregularities in the rotation rate of the Earth on time scales of a few days to several years. These are closely related to changes in the Earth's atmosphere. When the atmosphere is unusually warm, it expands, increasing the moment of inertia of the whole planet, which must then slow down to conserve angular momentum. Similarly a condensation of atmospheric water vapour on to the polar ice caps can reduce the moment of inertia and the rotation speeds up. The net result is that the length of the day can change irregularly by as much as 3 ms.

Weighing the Earth

A spacecraft from another civilisation, landing on Earth to obtain a sample of the surface, would most probably fall into the ocean, or hit an ice cap, or find soil and vegetation, all of which conceal the underlying rocky crust of granite and basalt. Our deepest drillholes reach no more than 12 km into this layer, but we know that it extends up to 40 km below our land masses. The crust is much thinner under the ocean floor, where it may be only 5 km thick.

The average density of the crust is about 3 g cm^{-3}, but the average density of the whole Earth is 5.5 g cm^{-3}. The deep interior of the Earth must therefore be radically different from the material we find in the surface layers and by deep drilling. How do we know the mean density, which leads to such an important result? The measurement which gives the density is the weighing of the whole Earth.

The basis of this measurement is to compare the gravitational attraction between two known masses with the gravitational attraction between one known mass and the Earth, i.e. its weight. The first measurement of this kind was made by the Astronomer Royal, Neville Maskelyne, in 1774, using an isolated mountain, Schiehallion, in Perthshire. On the northern and southern flanks of this 1000 m mountain Maskelyne established temporary observatories in which he measured the deflection from the vertical, caused by the mountain mass, of a plumb line. The deflection from the vertical was measured by observing the angular distance of a star from the zenith at transit as defined by the plumb line. The mass of the mountain was determined by surveying and by measuring the density of the rock. Maskelyne made 337 observations of 43 stars passing near the zenith at Schiehallion from which he concluded that the sum of the two contrary attractions of the mountain made a difference of 11.6 arc sec in the angular deflection of the plumb line. That is, the mountain caused a deflection from the vertical of 5.8 arc sec on a plumb line. From these measurements the density of the Earth was evaluated with an accuracy claimed to be 10%. The experiment incidentally provided a major test of the Newtonian gravitational theory – that is, the mountain and the plumb bob showed a measurable attraction to one another. In 1982, in celebration of the 250th anniversary of Maskelyne's birth, the Royal Society and the Royal Astronomical Society set up a plaque near the base of Schiehallion to commemorate this historic experiment.

In modern measurements the two attracting masses are very much smaller. A classic measurement by Boys in 1895 used two gold spheres 5 mm in diameter, each weighing only 1.3 g, suspended from a beam hung from a thin quartz fibre. These two spheres were gravitationally attracted by two lead spheres weighing 7 kg each, causing the fibre to twist proportional to the force. The result of such experiments gives the gravitational constant G, and the mass of the Earth (5.976×10^{24} kg).

The deep interior of the Earth

The most detailed information about the structure of the Earth's interior has been obtained from the study of the paths and velocities of seismic waves arising from earthquakes. From these observations there appears to be a series of steps in density, forming identifiable shells around a central core. The surface crust, with density 2.9 g cm^{-3}, is made of granite under the continents and basalt under the oceans. At the base of the crust there is a sharp discontinuity and the density rises to 3.3 g cm^{-3}. This is known as the Mohorovicic (or Moho) discontinuity after the man who discovered it in 1909. Ambitious plans to drill into the plane of the Moho discontinuity have been made but have not yet materialised.

The observed features of the Earth's crust and the results of geological and geophysical investigations can be explained in terms of continental 'plates' that are in constant motion relative to each other overlying a rocky mantle and a hot, molten, metallic core that is compressed solid at the centre.

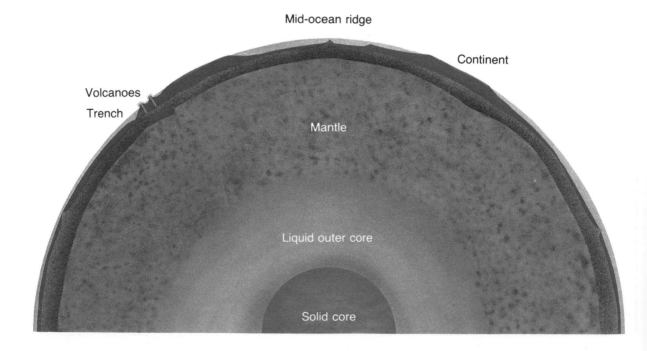

Mid-ocean ridge

Continent

Volcanoes

Trench

Mantle

Liquid outer core

Solid core

If this can be done in the future the composition of the mantle which lies beneath the crust will be determined by sampling. At present the mantle is thought to consist of rock rich in magnesium and iron silicates. At greater depth the mantle is divided by another discontinuity in density from 3.3 to 4.3 g cm^{-3} probably caused by the pressure of the rocks above giving rise to the higher density without significant change of chemical composition. At a depth of about 2900 km in the mantle another discontinuity increases the density to 10 g cm^{-3} and this marks the outer region of the Earth's *core*. The core is believed to be composed of iron and nickel oxides and to be divided into two zones. From the behaviour of the earthquake waves it is believed that the outer zone of the core is in a liquid condition and that this becomes solid at a depth of about 5100 km, indicated by another discontinuity in density. At this boundary between the outer and inner cores the density becomes 13.3 g cm^{-3}, and at the centre of the Earth where the depth is 6370 km the density has increased to 13.6 g cm^{-3}. The temperature at the centre of the Earth is believed to be about 4270 K; at the core–mantle interface it is about 3770 K, decreasing to about 1070 K in the upper mantle.

It is now widely accepted that the major surface features of the Earth have been determined by the movement of six mobile plates. These rigid slabs are formed of the crust of the Earth and part of the upper mantle. They are in continuous motion, and it is the movement of these plates relative to one another over hundreds of

The Earth's major tectonic plates.

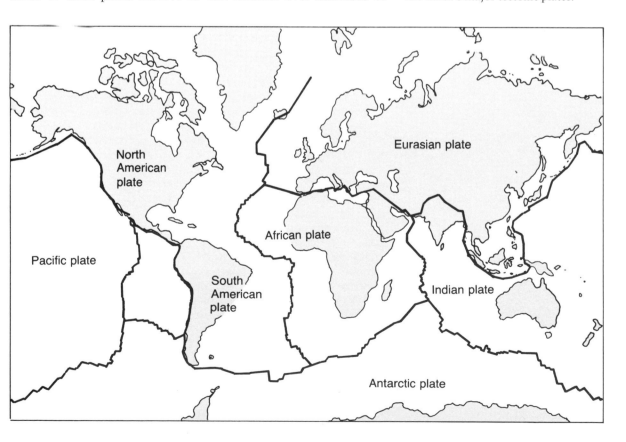

North American plate

Eurasian plate

African plate

Pacific plate

South American plate

Indian plate

Antarctic plate

millions of years that is believed to have determined the present distribution of the continents and oceans. In regions where two plates are moving apart the phenomenon of sea floor spreading is observed; for example, the Atlantic is opening up at the rate of 2 cm per year. In the regions of the Earth where two plates are sliding into one another the great mountain regions have been formed. Where the plates are sliding past one another *faults* are created – the San Andreas fault across California being a well known example. It is release of energy caused by the compression or snapping of stretched rock along these faults that gives rise to earthquakes. The study of the past and future shape of the continents and oceans based on the movements of these major plates is a fascinating subject. For example, the analysis of fossils and the study of the magnetic properties of rocks from different regions of the Earth provides the evidence that Europe and North America have been moving northwards for the last 100 million years at an average rate of 11 km per million years and that the drift rates for India and Australia have been 48 km per million years.

The Earth's magnetism and magnetosphere

If we suspend a common bar magnet by a thread it will come to rest with one end pointing towards the North Pole and the other end towards the South Pole. However, if we are navigators we have to take account of the fact that our magnet does not point exactly at the north and south geographic poles. The angle between the geographic and magnetic poles of the Earth – the magnetic declination – varies slowly. In 1970 it was 11.4° and it is increasing by 0.04° every ten years. The longitude of the north magnetic pole is 70.1° and is increasing by 0.07° every decade. It is as though there were a giant bar magnet aligned at an angle of 11.4° to the spin axis of the Earth. However, if we now mount a small magnet or compass needle so that it is pivoted horizontally we can measure the dip or inclination of the Earth's magnetic field, and we would discover that this simple analogy is not quite correct. By measuring the dip angle at different latitudes we would find that the lines of force are not vertical at the north and south magnetic poles as would be the case for a bar magnet – neither is the south dip pole opposite the north dip pole. At present the north dip pole is at 76°N, 101°W and the south dip pole is at 66°S, 140°E. Reference has already been made to the study of the magnetic properties of rocks in establishing the theory of continental drift. This is based on the assumption that over tens of thousands of years the geomagnetic axis averages out to the direction of the geographic axis. The remanent magnetism of rocks will have been frozen in at the time of their solidification and the lines of force will be those of the Earth's magnetic field at that epoch. Hence if the ages of the rocks are known their geographic position at the epoch of their solidification can be determined from these measurements of their remanent magnetism.

The origin of the Earth's magnetic field has been the subject of much discussion, particularly whether the Earth is a permanent magnet or whether the field is associated with its rotation – that it is some kind of dynamo. The second of these possibilities is now considered to be the case, and that the magnetic field is generated in the outer core where the molten iron would be a good conductor. The unsolved problem is the means by which an electric current is maintained in this core. One theory is that the outward flow of heat from the interior gives rise to convective motions and that the combined effect of these and the Earth's rotation is responsible.

The strength of the magnetic field varies from about 0.3 gauss (0.3 G) at the equator to 0.6 G at the poles. Although it is a simple matter to measure the field strength at any point, the detailed study is complicated because of the variations with time and position. Reliable measurements of the Earth's magnetic field have been made for over 300 years, and the longer term variations in the field (the secular variation) have recently been used to infer the nature of the changes occurring at the core–mantle boundary some 3000 km below the surface of the Earth. The variations of the intensity of the magnetic field have been recorded since about the middle of the 19th century. Before that time the measurements used were of the changes in the direction of the magnetic field dating back to the comprehensive observations of Edmond Halley in the 17th century. The results of this analysis were published in 1985 by Jeremy Bloxham and David Gubbins of the University of Cambridge. They conclude that there is diffusion of the magnetic field at the core–mantle boundary, implying that the flow of the fluid core is coupled to the mantle. Three possible plausible coupling mechanisms are suggested: it may either be thermal arising from temperature variations in the mantle which could drive the flow in the core; an electromagnetic coupling could arise from changes in the electrical conductivity of the mantle; or the coupling may be topographic caused by bumps and irregularities at the core–mantle boundary which could influence the flow in the core.

In recent years the detailed study of the magnetic field outside the Earth (the magnetosphere) has been made possible by equipment carried in Earth satellites and space probes. The magnetosphere is dominated by the solar wind, that is the outward flow of charged particles from the Sun, mainly electrons, protons and alpha-particles. At the distance of the Earth this solar wind has a velocity of some 400 km s^{-1}, and on the sunward side of the Earth this restricts the influence of the Earth's magnetic field to about ten Earth radii. On the dark side of the Earth the solar wind has the opposite effect, dragging the magnetic field lines so that their influence extends beyond the Moon. The major features of the magnetosphere are the zones of trapped particles discovered by van Allen in 1958 from the effects on Geiger counters carried in the first American Earth satellite. Two main van Allen belts have been delineated in which charged particles are trapped and spiral along the Earth's lines of force from pole to pole with periods from 0.1 to 3 s. The inner zone at a distance of about 1.5 Earth radii consists of trapped protons, and the outer zone at about five Earth radii consists of electrons and low-energy protons. Solar eruptions lead to sudden modifications of the configuration of the magnetosphere and to the terrestrial phenomena of magnetic storms and aurorae.

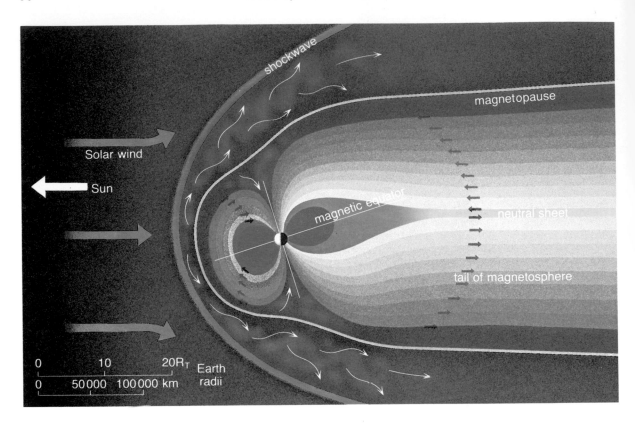

This schematic representation of the Earth's magnetosphere shows the shock wave that accompanies the Earth as it travels at supersonic speed through the solar wind, the magnetopause – the effective boundary of the magnetosphere – and its long tail away from the Sun. Electrically charged particles are guided along in the direction of the magnetic field. The Van Allen belts, in which charged particles are trapped, are shown in red and orange.

The atmosphere of the Earth

The plain statement that the air we breathe at the surface of the Earth comprises 78.08% by volume of nitrogen, 20.94% of oxygen and less than 1% of other gases gives no indication of the mechanism by which this balance has been established or how these gases, so essential to life, originated. We discuss these points in Chapters 9 and 21. Here we refer to some features of the structure of the atmosphere as we understand it today.

As we ascend above sea level by climbing a high mountain or by driving over a high mountain pass it is the decrease of pressure and density rather than any change in composition of the air that causes us difficulty in breathing. To a certain extent the human system can adjust itself to perform with reasonable normality up to altitudes of 3 to 6 km. However, a quick ascent to these levels – for example, by driving up the 4.25 km Pike's Peak in Colorado – makes one well aware of the sensitivity of the human body to a halving of the atmospheric pressure. At these levels also the average temperature has decreased by some 30 K. At the level of a jet aircraft flying at 12 km it would not be possible to survive without artificial pressurisation and heating. At these levels the atmospheric pressure is six or seven times less than that at sea level and the temperature will be 70 K lower than the average at sea level. In fact, at these altitudes in mid-latitudes, we are near the transition zone (the tropopause) between the troposphere and the stratosphere where there is a

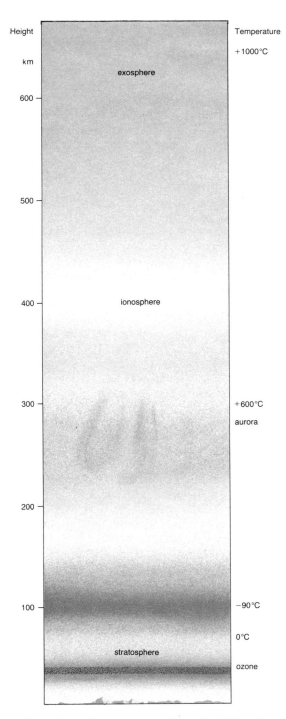

The main regions of the Earth's atmosphere. Eighty per cent of the mass of the atmosphere is concentrated in a thin layer at the base of the atmosphere, roughly equivalent in thickness to the skin on an apple.

temperature inversion. That is, as the altitude still further increases the temperature begins to rise. The height of the tropopause varies markedly with latitude. Over the poles the change occurs at about 8 km, but over the equator the tropopause is at an altitude of 17 km.

The relatively thin layer at the base of the atmosphere (the troposphere) contains 80% of the mass of the entire Earth's atmosphere. It is this region of the atmosphere in which the wind speeds increase with altitude and where there are appreciable vertical air movements that largely determine the weather conditions on Earth. Although ozone is present in the atmosphere to the extent of less than one part in ten million it is concentrated in a region of the high stratosphere at altitudes between about 20 and 50 km. The presence of this ozone is of great significance because it is the main atmospheric agent absorbing the intense ultraviolet radiation from the Sun. In recent years great concern has been expressed that the chlorofluorocarbon gases, commonly used in various types of household pressure sprays, have interacted with this ozone, allowing potentially damaging amounts of ultraviolet radiation to reach the Earth's surface, particularly over Antarctica. As it is, the ozone reactions and the absorption and scattering of the solar radiation in the troposphere reduce the intensity of the radiation by over one-half before it reaches the surface of the Earth.

The fraction of the incident solar radiation that does penetrate the atmosphere raises the temperature of the Earth's surface to a mean temperature of 287 K. Thus, whereas the peak intensity of the incident solar radiation is at the wavelength of visible light as expected from the solar temperature of 5800 K, the re-radiation from the low temperature Earth is at longer wavelengths (infrared). This radiation is absorbed by the carbon dioxide and water vapour in the troposphere. In this manner the constitution of the troposphere, and particularly the presence of carbon dioxide, determines the delicate heat balance that makes life possible on Earth. In this respect, as will be described in Chapter 9, the Earth is sharply distinguished from the neighbouring planets Venus and Mars.

Between altitudes of about 80 and 500 km there is a further important region of the atmosphere. This is the ionosphere, where the solar radiation (principally Lyman-alpha and X-rays) ionises the atmospheric constituents. At these levels the proportions of molecular nitrogen and oxygen decrease with altitude in favour of atomic oxygen. The density is so low that the equilibrium condition between the processes of recombination and ionisation results in a permanent ionised region, the density of the electrons and ions being markedly under solar control. Radio waves of long wavelength are reflected in the ionosphere, and until the advent of communication satellites this was the main process used in transoceanic radio communication. Because of the disturbance in the ionised regions caused by solar flares these long distance radio transmissions were frequently subject to fading and occasional complete blackouts. The radio waves used with the communication satellites are of sufficiently short wavelength to penetrate the ionised regions without scattering or absorption. Above the ionosphere at altitudes approaching 1000 km the pressure of the atmosphere has diminished to vanishingly small levels, and in these regions the magnetic field of the Earth begins to exert a dominant influence in the magnetospheric zones as already described.

8

The planets and their satellites

Seen from the distance of even the nearest star, the solar system might appear minute and compact. In reality, it is a vast, nearly empty space, populated with small bodies a great distance apart and insignificant in size compared with the Sun. Some idea of the sizes and distances can be gained from a model in the gardens outside the Jodrell Bank Visitor Centre: here a globe 1 m in diameter represents the Sun, and the planets are represented by spheres whose diameters and distances are scaled by the same factor of 1.4×10^9 times. Mercury then becomes a sphere with a diameter of 3 mm at a distance of 50 m; Earth, with a diameter of 1 cm, lies at 100 m; Jupiter, the largest planet, has a diameter of 11 cm, and lies at a distance of 545 m; Pluto, about the same size as Venus, lies much too far away for the model: it would be represented by a small pea at a distance of 4 km. The impression, as one walks through the model, is of open space.

Although the dynamics of the planetary system have long been understood, the accurate scale of the system has only recently been firmly established. Prior to the use of radar and of space probes, the best technique for establishing the scale was to observe the apparent position of one of our closest neighbours from two or more widely separated points on Earth, using the familiar technique of triangulation but with a baseline of thousands of kilometres. The closest neighbour was usually Venus, at a time when it was closest to the Earth. The planet could then only be seen when it passed across the solar disc, an event known as a transit. Such transits are rare because the orbits of Earth and Venus do not lie exactly in the same plane. The minor planet Eros proves to be a better subject, as it comes closer to Earth and also occasionally can be seen against the disc of the Sun. Heroic efforts are required for the observations; the voyage of Captain Cook in 1768–71 included a special visit to the island of Tahiti for the purpose of observing the transit of Venus.

Despite all such efforts, which should have produced an answer with accuracy of one part in 10^4, the various measurements were spread over ten times that range. In 1961 the first Soviet space probe was due to be launched to the planet Venus, and an accurate distance was urgently needed. Fortunately radar measurements became available while the probe was actually on its way, and mid-course corrections could be made. The scale of the solar system is now one of the most accurately known quantities.

A schematic representation of the solar system. The huge distances between the planets compared to their sizes make a scale drawing impossible.

Observing the planets

Apart from the simple identification of the brighter planets, and the more challenging hunt for Mercury, Uranus and Neptune, the amateur observer has little scope for exploring any details of our planetary system. It is, nevertheless, very rewarding to follow the phases of Venus and the Galilean satellites of Jupiter, both easily possible with binoculars, and the rings of Saturn make a glorious sight in a modest telescope. Mars and Jupiter both show changing surface features which a patient observer can follow. But terrestrially based observations have been largely superseded by the new era of space probes, and the spectacular pictures illustrating this chapter show details which were undreamed of before the epoch-making series of *Venera, Mariner* and *Voyager* spacecraft. The culmination was, and is, *Voyager 2*, which is making a grand tour of the giant outer planets; having visited Jupiter in 1979, Saturn in 1980 and Uranus in 1986, *Voyager 2* will visit Neptune in 1989, passing beyond the orbit of Pluto in 1990.

The physical nature of the planets

Table 8.1 shows immediately the distinctive categories of the inner planets, Mercury, Venus, Earth and Mars, and the outer planets Jupiter, Saturn, Uranus and Neptune. A third category comprises the minor planets and the satellites, which are more closely related to the inner, terrestrial planets than to the outer, giant planets.

The inner planets have densities not far different from the mean density of Earth i.e. 5.5 g cm^{-3}. Mars is appreciably less dense, at

3.9 g cm^{-3}, but it is still a solid planet made of rocks that are closely related to terrestrial rocks. The outer planets have densities closer to that of water, varying from 0.7 g cm^{-3} for Saturn to 1.66 g cm^{-3} for Neptune. There is no solid material at the surfaces of these planets; they are composed mainly of the gases hydrogen and helium, and their constitution is more closely related to that of the Sun rather than that of Earth.

Table 8.1 *The sizes and masses of the planets*

	Radius, km	Mass, kg
Mercury	2439	3.30×10^{23}
Venus	6052	4.87×10^{24}
Earth	6378	5.98×10^{24}
Mars	3397	6.42×10^{23}
Jupiter	71 400	1.90×10^{27}
Saturn	60 000	5.69×10^{26}
Uranus	26 000	8.70×10^{25}
Neptune	24 200	1.03×10^{26}
Pluto	1100-1700 (?)	6.6×10^{21} (?)

(?) Exact values unknown

The surfaces of Mercury and Mars have been thoroughly photographed by orbiting spacecraft. Venus is more difficult to explore because of its thick atmosphere, and only a few local photographs have been obtained from Russian probes landed on the surface. It is nevertheless possible to map the surface in great detail by radar, both from Earth and from orbiting spacecraft, using a technique called synthetic aperture radar. (The same technique was sub-

The cratered surface of Mercury as a photomosaic of images from the Mariner 10 probe of 1974–75.

sequently applied on the Earth satellite *Seasat*, which made radar images of the sea surface.) All these solid, rocky planets, including the Earth and the Moon, have been bombarded and battered with meteorites. On some planets, notably the Earth, there is little remaining trace of the bombardment, since the surface has been greatly modified by the large scale movements of the crust and by weathering of the surface features. On others, notably Mars and the Moon, large areas have been covered with lava from volcanoes. The surfaces of Mercury and Galilean satellite Callisto are covered with craters, many of them 100 km across.

The heavily cratered surfaces are all old, probably dating back to a period soon after the formation of the planetary system. The meteorite bombardment was the last phase of the accretion process, in which the planets were sweeping up material from the remains of the original disc concentration. Impacts of large meteorites are now comparatively rare, but they do occur in recorded history on Earth. They have obviously occurred comparatively recently on the other planets, since the areas which appear smoothly covered with lava or ice show a more scattered distribution of smaller craters.

The volcanoes which produced the lava flows are now inactive, except on the Galilean satellite Io. This satellite is trapped in the three-body resonance with its companions Europa and Ganymede, and it is forced to move through the very strong gravitational field

A radar image of part of the Ishtar Terra region of the surface of Venus, from the Soviet Venera 15 and 16 orbiters that operated between 1983 and 1985. The resolution of about one or two kilometres shows up grooved structures around a smooth plane containing what appear to be volcanic craters.

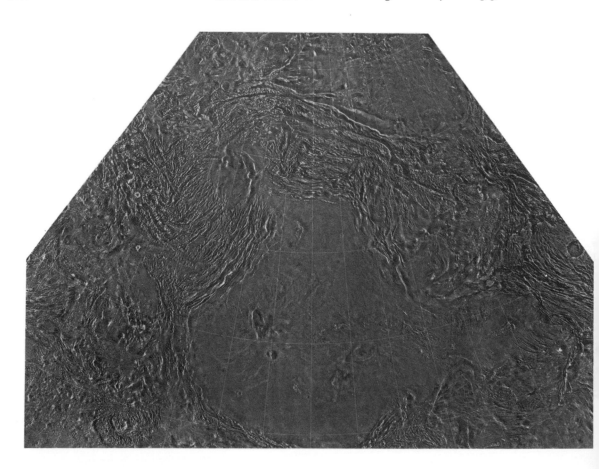

of Jupiter in a way that stresses the interior and dissipates energy. No other planet or satellite has such a source of internal heating. There are at least eight active volcanoes on Io, and its surface is covered with lava flows. The volcanoes also spew out sulphur and sulphur dioxide, which solidify on the surface, colouring it red, yellow and black. There are no meteor craters on Io; even the largest craters would have been obliterated by volcanic activity within the last few million years.

Our nearest neighbours, Venus and Mars, have naturally been the subject of speculation about the possibility that life could exist elsewhere than on Earth. The requirements seem to be an atmosphere, a moderate temperature, and water. Venus has a poisonous atmosphere and a surface temperature high enough to melt lead. Mars is a more likely haven for life. We discuss its atmosphere in Chapter 9, but we note here that its surface does show remarkable indications that there was at one time water flowing on the surface. The traces of rivers, complete with flow marks round large rocks, are very different from the famous 'canals' which were described by Schiaparelli and others in the 19th century. There is no trace of any activity by intelligent beings, or indeed any trace of biological activity of any kind. Any water that remains on the planet is now frozen, either under the surface or at the polar ice caps; even these ice caps are mainly solid carbon dioxide rather than water ice.

Mountains on Earth are mainly the result of continental drift. When large land masses move toward one another, the surface crumples into long ridges such as the Andes and the Himalayas. Tectonic movement has affected the surfaces of the other solid planets to a varying degree. On Mercury there are surface cracks, forming cliffs 100 km or more in length and some hundreds of metres high; these probably result from a general shrinkage of the surface. Most of Venus is flat, within an altitude range of about 500 m, but there are mountains up to 10 km high; these may be mainly volcanic rather than tectonic.

The great volcanoes on Mars are grouped around a high dome, Tharsis, which is the most evident result of tectonic activity on any planet except Earth. The dome probably formed as the planet

Although there is no liquid water on the planet Mars now, the existence of winding valleys such as the one shown here in a Viking orbiter photomosaic, strongly suggest that water once flowed in abundance on the Martian surface.

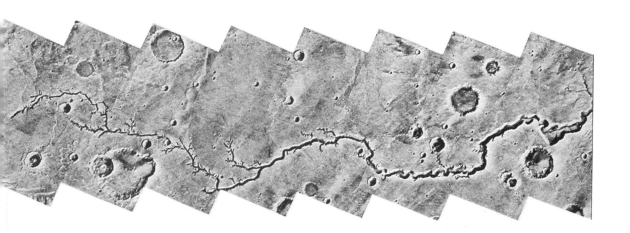

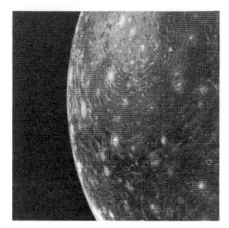

Part of the surface of the Jovian satellite, Callisto, photographed by the Voyager 1 probe in 1979. The outer ring of the huge impact basin visible in this image is 2600 kilometres across.

The extraordinary appearance of Io's surface is the result of sulphurous deposits spewed out from volcanoes that are kept constantly active by the internal heat generated through the gravitational interaction between the moon and its parent planet, Jupiter.

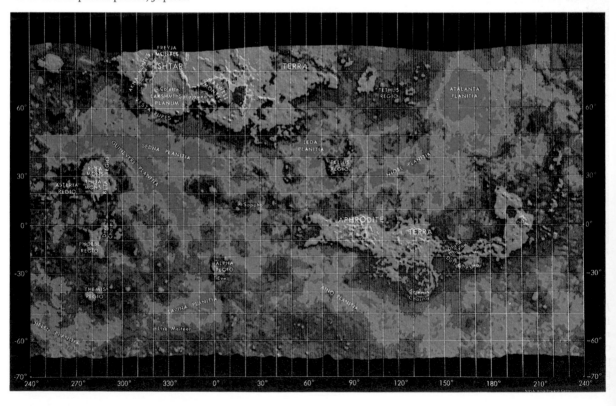

Left Thick layers of opaque cloud permanently conceal the surface features of Venus in visible light so planetary scientists have employed radar techniques to map the planet. Data from the Pioneer Venus probe of 1978 showed that sixty percent of the surface consists of a huge rolling plain, here coded light blue and blue-green. The highlands, coloured green, yellow and red, cover twenty-four percent while the remaining sixteen percent (dark blue) lies below the mean surface level.

This detailed image of Mars has been constructed from 102 separate photographs taken by Viking orbiter spacecraft. The long Valles Marineris canyon system and the three huge extinct volcanoes of the Tharsis region are prominent features.

solidified, as a result of an asymmetric distribution of mass. The volcanoes were a secondary effect of the strains in the surface material as the cooling proceeded. The evidence today is that the crust of Mars is fixed and there there are no large scale plate tectonic movements as on Earth.

The outer planets

Jupiter, Saturn, Uranus and Neptune form an obvious group of massive gaseous planets. Pluto is the misfit in the sequence of outer planets; it may be a former satellite of Neptune, thrown into its own orbit by an early gravitational disturbance, such as would be caused by a passing star. The two remaining satellites of Neptune, Triton and Nereid, are certainly in very unusual orbits; Triton is in a retrograde orbit, and Nereid is in an orbit inclined at 27° to the ecliptic. Pluto, and its recently discovered close companion Charon, are therefore to be classed with the satellites rather than with the other planets.

The outer planets are massive, and at the same time have a low density compared with Earth (see Table 8.2). Their outer parts are 99% hydrogen and helium, but their cores probably contain a mixture of heavy elements similar to that found in the inner planets. Uranus and Neptune show traces of methane, ammonia and water ice at their surfaces. The important factor here is the mass, which results in a gravitational attraction strong enough to retain the light gases and concentrate the heavier material in a hidden, central core. Another factor in the formation process may have been temperature, since the inner part of the solar system may have been warmed by radiation from the Sun during its formation.

Table 8.2 *Comparison of mass and density of the outer planets with that of Earth*

	Mass (Earth=1)	Density (Earth=1)	Rotation period (h)
Jupiter	318	0.24	9.9
Saturn	95	0.12	10.2
Uranus	14.5	0.22	17.24*
Neptune	17.2	0.30	18.0(?)*

*The rotation periods of Uranus and Neptune have never been determined satisfactorily from Earth based measurements, and values varying by a factor of two appear in the literature. The rotation period given here for Uranus was determined during the flyby of *Voyager 2* (January–February 1986) from radio astronomical and magnetic data (*Nature* 322, p. 42, 1986)
(?) Exact values unknown

The first three of these giant planets have rings in orbit round them. Saturn's rings may be seen through a small telescope, but the close-up views of *Voyager 1* and *Voyager 2* reveal an astonishingly detailed structure. The rings are made up of pebble-sized objects, following closely defined tracks like the grooves of a gramophone record. The disc is very thin, less than 1 km thick. Every 14 years it is seen edge-on, and it then virtually disappears. It extends from near the surface of Saturn, where the orbital period is less than 8 h, to 500 000 km out, where the orbital period is over 14 h. This range of periods spans the rotation period of Saturn, which is 10 h.

Saturn has at least 23 satellites, most of which are located outside the ring system. Large satellites cannot orbit close to such a massive body, as tidal forces would disrupt them. The limit of distance, known as the Roche limit, is about 140 000 km for Saturn, which is at the outer limit of the brighter part of the ring system. Inside this limit there are probably many smaller bodies, the larger of which influence the structure of the rings.

The satellites in the outer parts of the rings are obviously interacting with the rings and with each other. Three are in the same orbit; Dione, at a distance of 377 000 km, has two companions, one leading and the other lagging 60° away in orbit.

Jupiter's ring system is more tenuous; it has only been seen nearly edge-on by the Voyager probes. It is, however, sufficiently dense to be observable by infrared telescopes, which are particularly sensitive to radiation from dust particles. The ring system round Uranus was discovered in 1977 during the observations of the occultation of a bright star by the planet. The rings are narrow and well separated. They are very dark; their reflectivity is approximately equal to that of charcoal.

The asteroids, or minor planets

Taking a sky survey photographic plate, many degrees across and requiring an exposure of up to an hour, is a risky business. Many such plates carry a streak of light from a meteor or an artificial satellite, and some show the slower-moving streak of light from an asteroid, or minor planet. There are so many of these that some astronomers refer to them as the 'vermin of the skies'. Thousands of them have precisely determined orbits, and many more are discovered every year. They are, however, very minor members of the solar system, with a total mass only about 1% that of the Earth. The largest, Ceres, is about 1000 km in diameter, and itself accounts for half of the total mass.

Most asteroids are in a belt of orbits at distances between 2.5 and 3.5 AU from the Sun. Beyond this main belt there is an interesting group, the Trojans, in the same orbit as Jupiter at 5 AU. In a similar dynamical configuration as Dione and its two companions in orbit round Saturn, the Trojans occupy two stable positions leading and lagging Jupiter by 60° in its orbit.

Other asteroids are in more elliptical orbits. Members of the Apollo group, which cross the Earth's orbit, cannot survive for long in comparison with the lifetime of the whole system since they are certain to collide with one of the inner planets. They were probably thrown into their present orbits by a near collision with Jupiter.

Sunlight reflected from the asteroids shows that they have various compositions, falling mainly into the same groups as the meteorites observed to fall on Earth. Some are rocky, and others are metallic. They are presumably made up of material from the origi-

A close-up of part of Saturn's ring system, photographed by Voyager 2 on 25th August 1981 and shown in false colours, which indicate that different sections of the rings have different compositions.

nal solar nebula, but without the gases which are retained by the planets, and without the modifications which volcanoes, tectonics and the weather so obviously cause on Earth. The capture and return to Earth of a piece of this original material now seems to be a feasible mission for a spacecraft.

Opposite The orbits of most asteroids lie within a belt between those of Mars and Jupiter. Some are known that have much more elliptical orbits and pass close to the Earth and Sun. The Trojan asteroids are groups sharing Jupiter's orbit but confined to gravitationally stable positions 60° either side of the giant planet.

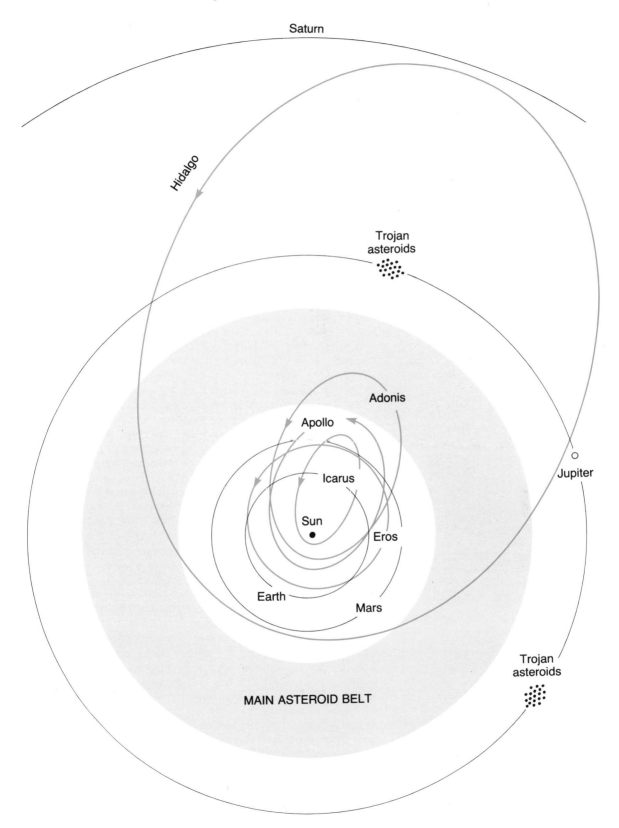

The origin of the solar system

When the infrared satellite telescope IRAS made the first survey of the whole sky from above the atmosphere, the bright star Vega was used as a calibration of flux. Vega, however, turned out to be a bad choice; it has far more long wavelength infrared radiation than the star itself can produce. It is, in fact, surrounded by a cold dusty nebula such as may have been the origin of our planetary system.

Many other stars are now known to have similar cool nebulae round them, although it is not yet possible to say that any material has aggregated into substantial planets. The existence of these nebulae nevertheless supports the theory of the solar nebula to which we have referred throughout this chapter. The theory is necessarily speculative. We have only one solar system as example, and inside this solar system we have an astonishing variety and far too much detail for any simple theory to encompass. Why, for example, are the four terrestrial planets nearest the Sun – Mercury, Venus, Earth, Mars – over 200 times less massive in total than the outer planets Jupiter, Saturn, Uranus and Neptune? Is the low density of the outer planets solely due to their retention of the light gases hydrogen and helium, or was the nebula from which they formed already differentiating into heavy and light elements? What was the process which decided the spacings between the planetary orbits?

On the whole, contemporary opinion supports the nebular hypothesis of the type which was first put forward by Laplace in 1796. The solar nebula is usually regarded as an isolated evolving body, but in some versions of the theory the Sun acquired a nebula after its own formation, during an encounter with an interstellar gas cloud. In another version, now discarded, the nebula was drawn out from the Sun by a close encounter with a passing star.

The original condensing nebula may have resembled one of the dark globules which can be observed in the spiral arms of the Milky Way, as for example in the Orion Nebula. These globules contain the 10^{57} atoms required to form a star, but the process of further condensation is still mysterious. Any theory of the later stages of condensation into a disc, and into planets, must, for example, explain how the angular momentum of a slowly rotating nebula becomes distributed among the various bodies; why do most planets rotate with a period of about a day or less, and why does the Sun, with 99.9% of the mass, contain only 2% of the total angular momentum of the whole system?

A final and most fundamental question concerns the abundance of the elements. Where did all the heavy material come from – heavier, that is, than the original hydrogen and helium of the early Universe? The present view is that the solar nebula condensed from the clouds of material resulting from the earlier formation and evolution of massive stars, whose internal furnaces and supernova explosions were responsible for the process of nucleogenesis. Without these processes there would be no solid planets, and no astronomers to speculate on their origin.

9

Atmospheres and climates of the planets

The surface of the Earth is the only place (as far as we know) where we can live and breathe normally. We begin to gasp for breath on a high mountain. Men have survived at 8.8 km (29 000 ft) on Mt. Everest but only for brief periods after a long acclimatisation and they often require a supply of oxygen. At greater altitudes we could not survive without an artificially created environment. In a jet air-craft at 12 km (40 000 ft) the cabin has to be pressurised and heated. Not only the pressure and temperature, but also the con-stitution of the air we breathe, has to be correct with the right mixture of gases, especially oxygen and nitrogen. The question as to why such a narrow region close to the Earth is conducive to normal living is very important. We can specify the conditions and we understand fairly well how the balance is maintained, but do we know how these conditions arose in the first instance? Could we live on any other planet?

The inner planets, Mercury, Venus, Earth and Mars, are all solid bodies made of similar material. Their atmospheres are, however, totally different from one another. Two main factors determined the evolution of their atmospheres from the gas of the condensing solar nebula: their mass and their distance from the Sun. Distance determines the strength of solar radiation; Mercury receives six times as much energy per unit area as does Earth, and its sunlit side becomes extremely hot. As it spins very slowly, the other side becomes extremely cold. We could not live on Mercury with a temperature varying from 670 K in the daytime to 100 K at night.

A planetary atmosphere is kept in place by gravity; it follows that the more massive planets are more likely to have an atmosphere. Cold heavy gases are easier to retain than hot light gases. But Mercury and the Moon have retained none at all, in common with most of the planetary satellites. Mars is colder than Earth; it there-fore has an atmosphere despite its lower mass. Venus is closer to the Sun and slightly less massive than Earth; nevertheless, it has a very dense atmosphere, and we need to know why.

Venus, Earth and Mars have lost practically all of their original hydrogen and helium, which is still to be found in the heavier, outer planets. The gases which would be expected to remain are carbon dioxide, nitrogen and water vapour. Note that oxygen is not expected, since it would have combined with carbon and with hydrogen in the early stages of planetary condensation. Mars is the simplest example; its atmosphere is 95% carbon dioxide, the rest being nitrogen with a trace of water and other gases such as argon. It is too cold for water vapour, although the surface markings show

Clouds of water ice gather in this canyon on Mars as the Sun rises. The picture is reconstructed from three black and white images taken through coloured filters by the Viking 1 orbiter in 1976.

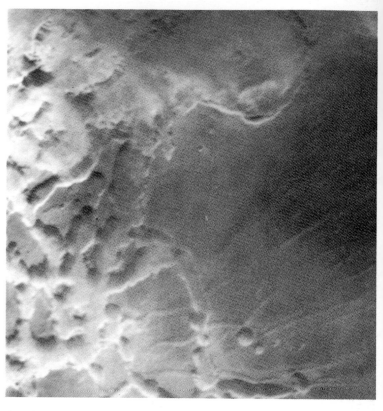

that liquid water once existed on Mars. It may now be condensed under the surface as permafrost.

Venus also has a predominantly carbon-dioxide atmosphere, but its temperature is high and the absence of water vapour cannot be explained by condensation into oceans and frost. In fact, Venus is so hot that water has probably been dissociated into hydrogen and oxygen which have escaped. The escape of oxygen would, however, be rather slow, lasting over most of the lifetime of the planet.

We see therefore an immense contrast between the atmospheres of Venus and Earth. Obviously Earth has retained enough water to fill the oceans; less obviously it has also retained its carbon dioxide, which was dissolved in the oceans, and is now retained in the form of carbonate rocks. Both these factors are strongly dependent on temperature, and we can see that Earth is at a critical distance from the Sun, where the temperature is between the boiling and freezing points of water. Venus is too close, and therefore too hot; Mars is too far away, and too cold to form oceans. On Venus the heavy gas carbon dioxide could neither escape nor dissolve in water, as it did on Earth. On the contrary, it was probably enhanced by volcanic activity.

The vital difference between the atmospheres of Earth and the other planets is the existence of oxygen in our atmosphere. Some oxygen can be expected in the original mix of gases in the condensing solar nebula, but this mixture alone would produce a primarily hydrogen and helium atmosphere, as in the giant planets. How did our planet acquire so much oxygen?

The evolution of the Earth's atmosphere must have been a complex process. There are indications that the atmosphere did not originate directly from the solar nebula. Most of the original gas must have been lost and replaced by outgassing from the solid planet. An indication of this is provided by the heavy gases argon and neon, which are unlikely to be lost but which are nevertheless very rare compared with the abundances of other heavy elements on Earth. Another remarkable indication is the absence of methane and ammonia which have formed in other planetary atmospheres from the material of the solar nebula.

We conclude that the Earth's atmosphere is the result of a secondary event. Even the existing water and nitrogen could have

The Earth as it appeared from space to the Apollo 17 astronauts. The continents of Africa and Antarctica can be dimly discerned but it is swirling white clouds and oceans that dominate the image.

originated from within the Earth, since volcanic activity during a few billion years could provide all the water in the oceans and all the nitrogen in the atmosphere. Some oxygen can be produced by the photo-dissociation of water, but the present level of oxygen is entirely due to the process of photo-synthesis in plants. The critical event was the beginning of life on Earth, about three billion years ago. Among the conditions which made this possible was the existence of enough oxygen for an ozone layer to form in the atmosphere, providing protection from the solar ultraviolet rays just as it does today. No other planet has produced more than a trace of oxygen, and no other planet has sustained life.

The climates of the inner planets

The temperatures of the planets are mainly determined by their distances from the Sun. There is a balance between absorbed solar energy and re-radiated heat, best measured in the infrared spectrum. The balance depends on the way solar radiation penetrates the atmosphere, and on the transparency of the atmosphere to the re-radiated infrared. On Mars, with a thin atmosphere, the balance is usually determined only on the surface; there are, however, occasional dust storms which blanket the infrared radiation and raise the temperature locally by as much as 50 K. On Venus there is a much greater effect, which raises the whole surface temperature to over 720 K, greater than the peak daytime temperature on Mercury. This is due to the carbon-dioxide atmosphere, which is relatively transparent to visible light and opaque to infrared.

There is a similar 'greenhouse' effect at work on Earth, but the more complex atmosphere limits the temperature rise to about 33 K only. The rise is sensitive to the amount of carbon dioxide in the atmosphere. Human activity is currently increasing the proportion of carbon dioxide by the burning of fossil fuel, and an anxious watch is being kept by meteorologists for any global increase in temperature.

Meteorologists have also been deeply involved in understanding the dynamic behaviour of the planetary atmospheres. Mars is the closest to Earth in its behaviour, even though it is too cold for water vapour to take any important part. There is a general circulation of the atmosphere, arising from the greater heating at the equator. There are clouds, probably of frozen carbon dioxide and some water ice; these clouds show some familiar behaviour in the wind patterns, such as the waves to be seen in the lee of mountains. The meteorology of Mars is very variable. During the day the temperature can range from 170 K to 270 K. There are also marked seasonal changes through the Martian year, which lasts through two terrestrial years.

Venus is permanently covered with cloud; the very bright planet on which we so frequently remark is in fact a cloud containing an invisible planet. The clouds are high in the atmosphere, about 60 km above the surface. They are composed of liquid drops, like an aerosol, probably of sulphuric acid. The atmospheric temperature and pressure both increase steadily downwards towards the surface, in contrast to the terrestrial atmosphere, where a trough of

In ultraviolet light, subtle patterns in Venus' couds are revealed, as in this picture taken by the Pioneer Venus orbiter in 1979. The cloud tops swirl round the planet in only four days, moving from the equator to the pole at the same time to create the characteristic V-shaped pattern.

low temperature is found at the tropopause. At the surface, the pressure on Venus reaches over 90 atmospheres, i.e. 90 times the terrestrial surface pressure. This is due to the carbon dioxide, which on Earth is conveniently tucked away in the carbonate rocks.

The dense atmosphere of Venus has well-established and stable wind patterns. The rotation, and the surface velocities, are so slow that Coriolis forces are negligible, and none of the familiar cyclonic and anti-cyclonic patterns appear. Instead there is a very high wind speed in an equatorial belt, like a very fast jet stream, with lower velocities at higher latitudes.

If by some artificial means we could manage to survive on Venus then, apart from the barrenness of the surface, we would find some startling effects. The high density of the atmosphere refracts a ray of light through 180°. Of course this refraction also occurs through the Earth's atmosphere but it is very small. On Venus, however, the effect would be so marked that we could see the Sun setting with our backs turned to it. Our view would not be of the globe of the Sun as we see it on Earth but of a great band of light stretching around the horizon. The normal horizon of the planet would vanish to be replaced by a series of curved bands of light which would give the impression that we were standing inside a giant spherical dome.

The giant planets

The terrestrial planets are characterised by having a solid surface, and, in the case of Venus, Earth and Mars, an atmosphere. Our view of the giant planets whether from Earth or from spacecraft is only of the outer cloud layers, and we have no direct measurements of conditions beneath the clouds.

Finely detailed patterns in the clouds of Jupiter, including the famous Great Red Spot, are revealed in Voyager pictures such as this one taken on 1st February 1979 from 32.7 million kilometres away.

Jupiter is by far the best explored of the giant planets, but there is good reason to believe that Saturn and probably Uranus and Neptune share many of the characteristics of its atmosphere. It is rich in phenomena which are new to atmospheric physics and chemistry; their study is valuable also to meteorologists, albeit as a contrast to the behaviour of the terrestrial atmosphere.

From an analysis of the ultraviolet, visible and infrared radiation observed from Earth and from the *Pioneer* and *Voyager* space probes, it was found that the outer 2000 km layer of the Jovian atmosphere is hydrogen at a temperature of 1500 K. The reflected sunlight comes from clouds which have a high reflectivity; they are probably composed of frozen ammonia, but there are absorption

The low contrast and lack of detail in the cloud patterns of Saturn, particularly in comparison with Jupiter, are due to haze. This picture was constructed from a mosaic of images taken by Voyager 1 on the 10th November 1980 from a distance of 18 million kilometres.

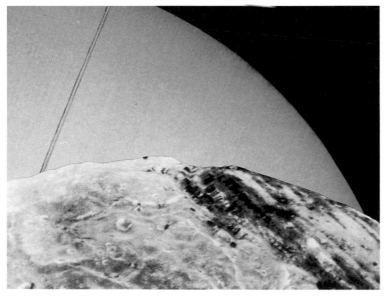

A montage of Voyager 2 images, obtained in January 1986, shows Uranus at a distance of 105 thousand kilometres, overlaid with an artist's conception of the planet's rings, and with the moon Miranda in the foreground.

bands in the spectrum of the reflected light indicating the presence of methane and molecular hydrogen. Helium is probably also present, but it does not produce absorption bands in the visible spectrum. The clouds are coloured variously and spectacularly, mainly yellowish in hue: this is probably due to various forms of sulphur condensed on to the otherwise white ice grains in the clouds.

The atmospheric circulation in Jupiter is dramatic. There are zones in latitude with persistent east–west winds, reversed in alternate zones; these may be seen from terrestrial observations as bands of colour. The long persistence of this wind pattern is to be expected, as the atmosphere is so massive that the rapid changes typical of the terrestrial atmosphere are impossible. There are, however, variable features which appear at the boundaries of the zones which are more reminiscent of the familiar cyclonic wind patterns on Earth. A major feature of this kind is the Great Red Spot, which has persisted for several centuries. It is an anti-cyclonic pattern of winds, outlining a cold area high in the atmosphere. The Great Red Spot is not attached to any other fixed feature; it appears to be a natural and semi-permanent feature of the parallel bands or zones of winds. Around it there can often be seen other less persistent markings, particularly some smaller white spots and some dramatic turbulent eddies.

The source of the zonal circulation is probably deep in the liquid interior of Jupiter. One theory is that the planet is composed of concentric cylindrical regions, whose ends appear on the surface in both hemispheres. These cylinders are forced into differential rotation by thermal energy. Some support for this is given by the observation that Jupiter radiates more energy than it receives, so that it must have an internal source of energy about which we can only speculate. Without this extra source of energy, Jupiter's temperature would be 105 K; the observed temperature is 125 K. The very vigorous circulation of Jupiter's atmosphere ensures a fairly even temperature over the whole planet; the difference between temperatures at the pole and at the equator is only a few degrees. Whether or not the internal source of energy is responsible for the zonal winds, it will certainly also induce some more local convection. Rapid rotation also means that Coriolis force is important, as can be seen from the circulation patterns round the Great Red Spot.

The atmosphere of Saturn follows the same general pattern as on Jupiter. There is a large thin cloud of hydrogen extending more than 10^6 km from the visible planet, and there is a zonal system of winds. Some oval features, like Jupiter's Great Red Spot but less persistent, can be seen from Earth. The band of wind at the equator is a jet stream with the very high velocity of 500 m s^{-1} (over 1000 mph).

Saturn also has an internal source of heat; as on Jupiter, this is detected as a higher temperature than expected from the simple balance between solar energy and re-radiated infrared.

Uranus and Neptune are less massive than Jupiter and Saturn, but they are colder and they are therefore able to retain a similar atmosphere of light gases. They are both rapid rotators, and the general atmospheric circulation can be expected to be similar to that on Jupiter. There are, however, no appreciable sources of internal energy on either Uranus or Neptune. Atmospheric circulation on Uranus will be particularly interesting to investigate; the rotation axis lies in the equatorial plane, so that one pole faces the Sun for several years.

Radio waves from the planets

The best estimates of the temperatures of the planets are derived from measurements of their infrared radiation. Thermal radiation can also be received in the short-wavelength part of the radio spectrum. For Mars the radio and infrared radiation both originate on the surface, but for Venus the radio emission originates from below the clouds, deeper down than the visible and infrared radiating regions. Similarly, in the giant planets the radio measurements give us the temperature of regions at a greater depth, and therefore at a higher temperature, than the infrared measurements. On Saturn, for example, the radio temperature is 250 K, although the coldest part of the atmosphere higher up is at 85 K.

There are, however, some far more spectacular radio emissions from the giant planets, and especially from Jupiter. These are associated with the general magnetic fields of the planets. Earth itself provides examples of radio noises generated in ionised gas trapped in the field lines of the dipole magnetic field. On Jupiter, this region of magnetic field and ionised gas, known as the magnetosphere, is an intense source of radio waves. It is fed with ionised gas by the volcanoes of the satellite Io. As a result of this surprising link between the planet and its satellite, the intensity of the radio waves varies both with the rotation of Jupiter and the revolution of Io.

The magnetic field of Jupiter originates deep in its interior, and is probably attached to a solid rocky core. If so, the rotation period measured by the radio waves must be the best definable for the planet. It is 9 h 55.4 m, which is close to the rotation period for clouds near the equator, but considerably faster than that at higher latitudes.

Life on other planets

The delicate balance which has resulted in tolerable conditions for life on Earth appears the more precarious as we learn more about the other planets. Our investigations can only encourage a degree of caution in any discussion of the possible presence of life forms on planetary systems that may exist elsewhere in the Universe. A stable habitable atmosphere can only exist on a planet in a very narrow zone from the parent star. Whether such a zone exists depends critically on the relation between the temperature of the star, the mass of a planet and its distance from the star. In the case of our own planetary system this zone is extremely narrow compared with the extent of the system. For a planet the size of Venus or Earth, the planet must be more distant than Venus for the reasons already given. On the other hand Mars is too small and distant. Nevertheless if a planet the size of Earth existed in a Martian orbit then the greater volcanic activity and the retention of a more substantial atmosphere would provide habitable conditions. Finally, any such conclusions must depend on the occurrence of whatever circumstance swept away the primeval atmospheres of the inner planets in their earliest history and allowed the secondary atmospheres formed from volcanic eruptions to prevail.

10

Comets

For more than 150 years the monthly meetings of the Royal Astronomical Society in London have been an important forum for the presentation of reports on astronomical research by professional astronomers. Many who were present at one of these regular meetings on 13 February 1981 will remember the occasion as unusual because an amateur astronomer was called upon to speak. Roy W. Panther of Walgrave near Northampton described how, during the wartime blackout early in 1943 when the skies were very dark, he saw a comet without any optical aid. After the war he

Roy Panther, an amateur astronomer from Northamptonshire, and the telescope with which he discovered a comet in 1980.

became an ardent comet hunter, initially using a 3 in refractor and later an 8 in f/4 reflector with a magnification of 35. The evening of Christmas Day 1980 was so clear that he decided to make a 'comet sweep' with this instrument. After about an hour he noticed an unusual 'misty round object' near the star Epsilon Lyrae. He had discovered a new comet known as Comet Panther 1980u.

In this way amateur observers make one of their most important contributions to astronomy, either by the 'recovery' (that is, the sighting of a comet discovered in earlier years) or by the discovery of a new comet that has never previously been observed. There is often a friendly rivalry between amateur and professionals. Amateurs like Mr Panther use small telescopes and sweep the sky looking for fuzzy objects that move amongst the stars. Professionals more often make discoveries on the routine photographic patrols of the sky. Compared with the amateurs they have the advantage of greater sensitivity, but on the other hand the amateur takes in much greater areas of sky during the nightly sweeps.

The nomenclature of comets

When the discovery of a new comet is first announced it is named after its discoverer followed by the year and a small letter indicating the order of discovery in that year. Thus, Comet Panther 1980u was

Comet Kohoutek photographed with the 48-inch Schmidt camera of the Palomar Mountain Observatory on the 12th January 1974.

the 21st comet to be discovered in 1980. Subsequently, when enough observations have been made so that the orbit of a comet can be computed, it is given a revised designation in terms of the year and order of perihelion passage. The year may not be the year of discovery, and the order is designated by a Roman numeral. This final designation is subject to the approval of the appropriate Commission of the International Astronomical Union. As an example, let us take one of the comets discovered by Fred Whipple of Harvard. He had been assigned the task of inspecting the quality of the photography and processing of the photographic plates taken for sky surveys by the Harvard College Observatory. Over a period of 12 years he scanned 70 000 glass negatives measuring 203 × 25 mm with a hand magnifying glass. Searching for cometry images was not his primary task; nevertheless, he discovered images of six comets on those plates. Only one of these comets – on a plate taken in 1933 – turned out to be of short period, and on its next appearance in 1941 it was the third comet to pass through perihelion in that year and received the designation 1941 III P/Whipple; the letter P indicates that it is a periodic comet. If an observer is fortunate enough to have discovered more than one comet this is indicated by the addition of an Arabic numeral after the name.

Table 10.1 *Discovery and recovery of Comet P/Whipple*

Year and order of discovery or recovery	Official nomenclature – year and order of perihelion passage
1933f	1933 IV
1940b	1941 III
1947g	1948 VI
1955d	1955 VIII
1962f	1963 II
1969c	1970 XIV
1977h	1978 VIII
1985h	1986 (perihelion 25 June)

The convention that a comet is named after its discoverer is broken in some special cases. The most notable example is that of Halley's comet. On the basis of Newton's law of gravitation Edmond Halley predicted the return in 1759 of a comet seen in 1531, 1607 and 1682. Historical research has revealed that records of this comet exist for the past 1000 years, and clearly the contemporary convention cannot apply in such a case. Detailed reference to this famous comet is made later.

The orbits of comets

In Chapter 5 reference was made to the Oort Cloud of comets. It is believed that near the limits of the solar system at a distance of about 50 000 AU from the Sun there exists an accumulation of some 10^{11} cometary nuclei. This cloud of potential comets is gravitationally bound to the Sun. Occasionally it is perturbed by a passing star which may result in the ejection of some of the comets, bringing them into a new orbit towards the Sun. Near perihelion

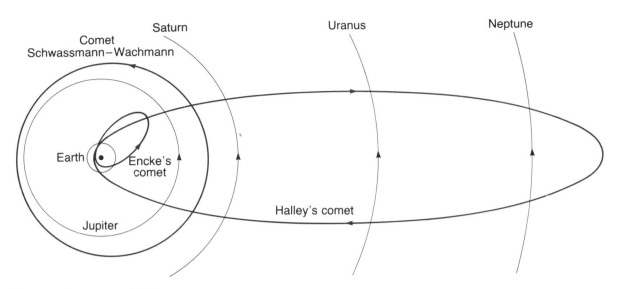

they may become visible from Earth, and given sufficient observations the orbit of the comet can be computed. Early in the 19th century new mathematical techniques were developed which facilitated the computation of cometary orbits from such observations. An authoritative contemporary (1979) catalogue of known cometary appearances throughout history has 1027 entries. When different returns of the same comet are allowed for, the list reduces to 658 different comets. The majority in this list (545) are long period comets, that is those with orbits that are computed to be nearly parabolic about the Sun with estimated orbital periods of several hundred to thousands of years. Of greater interest to terrestrial observers are the short period comets of which there are 113 in the list with periods of less than 200 years.

The short period comets were probably thrown into their present elliptical orbits by a close encounter with a planet, usually the heaviest planet, Jupiter. On a cosmic timescale they have short lives, since they can be disintegrated by collision or evaporated by solar radiation. The dusty remains of a comet may persist in the same elliptical orbit, possibly to be observed from Earth as a meteor shower. Thus relatively few of the short period comets can be observed regularly, as they return to the neighbourhood of the Sun and brighten up after years of invisibility. Seventy-two such comets have been observed at least twice. Of these the most famous is Halley's Comet with an orbital period of 76 years around the Sun. This orbital period is some ten times that of the average of the periodic comets and the fact that it has spent so much of its life at great distances from the Sun probably accounts for its longevity.

It is the uncertainty as to whether a given comet, after being invisible for years, will re-appear as predicted during the few months when it is near perihelion that makes the consistent comet watch so important and interesting. The number of comets that are 'recovered' – that is become visible again – in any year, or are newly discovered is a variable quantity. For example, in 1978 over 40 comets were under observation (an unusually large number). Of these, eleven were new discoveries, seven were 'recovered' and the

The periodic comets have become trapped in closed orbits within the solar system, probably through the gravitational influence of Jupiter. Many of their orbits are greatly elongated and inclined steeply to the main plane of the solar system.

remainder were still observable having been recovered the previous year. In 1979 only 26 comets were observable, of which seven were new discoveries. 1980 yielded a total of 29, of which ten were new discoveries. These various factors give a substantial measure of uncertainty to any list of comets. Also it happens that a comet claimed to be newly discovered may eventually transpire to be the return of a previously listed comet. A famous example of this occurred early in the 19th century. In 1818, at Marseilles, Pons discovered a new comet, and Encke, a pupil of the great mathematician Gauss, computed the orbit to be 3.3 years – incidentally one of the shortest periods of any known comet. Encke was able to show that this comet was identical with 'new' comets discovered by Mechain in 1786, by Caroline Herschel in 1795 and by Pons in 1805. Encke correctly predicted that the next perihelion passage would occur on 24 May 1822. With over 50 reappearances Comet Encke, as it became known, heads the list of known cometary returns of the short period comets.

The nature and origin of comets

To the terrestrial observer a comet appears as a fuzzy spot in the sky. As the comet approaches the Sun a diffuse tail can be seen aligned in a direction always away from the Sun. After perihelion passage the tail is generally most pronounced and often rapid variations in the structure of the comet can be seen. At successive appearances periodic comets often behave differently. In fact, this general evidence indicates that a comet is a fragile and unstable body.

The case of Biela's comet provides a dramatic illustration of this instability. In 1826 Biela discovered a new comet, but the computation of the orbit showed that this was the same comet that had been seen in 1772 and 1805. It was also realised that prominent meteor showers had occurred at the appropriate time of passage of the Earth through the cometary orbit. The comet was again seen in 1832 and 1846 but at this last return the comet was divided into two parts. These travelled side by side in space and appeared again

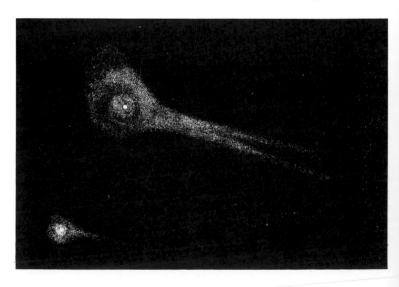

An engraving of Biela's comet, after drawings made by Struve, from a book published in Paris in 1870. The comet was discovered in 1826 and had a period of 6.6 years. When it reappeared in 1846 it had split in two. In 1852, both comets were seen again for the last time. However, in 1872, 1885 and 1892, when the Earth crossed the path of the lost comet, meteor swarms were observed.

at the 1852 return with the separation somewhat increased. After that return the twin comets were never seen again, but in 1872 when the Earth crossed the orbit 80 days behind the computed position of the lost comet there was a great meteor display. The next crossing of the orbit in 1879 was far from the computed cometary position and very few meteors were seen, but in 1885 another great meteor display occurred at the appropriate time of crossing the orbit. Since that time there have been no further meteor showers associated with the comet and it is presumed that the meteor swarm, being the remnants of the comet, passed close to Jupiter and was perturbed away from the Earth's orbit.

From the most detailed photographic and spectrographic evidence made from Earth it is conventional to specify three divisions in the structure of a comet. Near the head of the comet the *nucleus* has a stellar appearance in the bright comets. This is surrounded by the nebulous *coma* which merges into the *tail*. No part of a comet is self-luminous. It becomes visible as it approaches the Sun by the reflection of sunlight from the nucleus and by excitation of the molecules in the coma and tail. C_2, CH and CN are generally identified from the spectrum of the coma, and CO and N_2 persist in the tail. Several other molecules, notably OH and NH, have been identified. Most of these molecules are unstable when

This colour composite photograph of Comet Halley was made from three photographic plates exposed on the UK Schmidt Telescope in Australia on the night of the 12th March 1986. The lower filamentary tails are ionized gas; tails to the north are formed by dust being swept from the nucleus and shine by reflected sunlight. The telescope tracked the comet, which was moving relative to the stars, so the images of the stars look slightly trailed.

released from the coma near perihelion passage and it is the longer life molecules of CO and N_2 that are predominant in the tail.

If a comet crosses the line of sight from the Earth to the Sun it becomes invisible. In the 1910 transit of Halley's Comet no shadow of the comet could be seen on the solar disc although even the shadow of a solid body only 50 km across would have been visible as a small dot crossing the solar disc. Neither is the light of a star dimmed when it is occulted by the coma of a comet. During the close approach of the Pons–Winnecke Comet to the Earth in 1927 it was possible to observe the nucleus under high magnification and it was concluded that any solid body in the nucleus must be less than 400 m in diameter. It was therefore a very exciting moment in March 1986 when the spacecraft *Giotto* flew within 500 km of the nucleus of Comet Halley, and sent back the first picture of any cometary nucleus, which was seen to be an ellipsoid about 11 km long and 8 km across. The nucleus itself reflects very little sunlight, so that the Pons–Winnecke Comet may also have a similarly sized nucleus which was so dark as to be practically invisible in the close encounter of 1927.

Only the size of the nucleus of Halley's Comet is known, not its mass or even its composition. The material of the coma and the tail can, however, be investigated in some detail. This material blows off from the comet when it is close to the Sun; for the short period Comet Encke, which has already been observed for over 50 returns, the loss is at the rate of 200 kg s^{-1} while for brighter comets like Arend Roland (1951) the rate was more than 70 tonnes per second (t s^{-1}). For Comet Halley, which loses up to 40 t s^{-1}, each perihelion passage means the loss of a surface layer more than 1 m thick; 1000 more perihelion passages will see the comet sadly shrunken, unless it has already disrupted completely as did Biela's Comet.

The composition of the nucleus itself can only be deduced from the material of the coma and tail. Here there is some direct evidence from the dust particles from Comet Halley encountered by *Giotto*; these appear to be similar to meteoritic material which can be picked up on the surface of Earth, but with predominantly small sizes of order a few microns. The volatile constituents of the coma and tail can be observed both from spacecraft and from ground-based observatories. The spectrum of the light reflected or absorbed by the tail shows many molecular bands corresponding to simple combinations of the elements hydrogen, oxygen, nitrogen and carbon. Molecular hydrogen exists as a large cloud surrounding the tail, which contains molecules and radicals such as water vapour, ammonia, methyl, carbon dioxide, carbon monoxide and cyanogen. These are all materials which exist as solids (such as water ice and solid carbon dioxide) at sufficiently low temperatures.

The current view of the nucleus is the icy-conglomerate model proposed by Fred Whipple in 1950. Like a dirty snowball, where the snow melts or evaporates to leave a dirty outer surface, the volatile material of the solid nucleus evaporates and escapes, leaving a fluffy black exterior of meteoritic dust. This shell acts as a shield from solar radiation, and evaporation tends to occur in patches. The nucleus rotates (once every 2.2 days for Comet Halley), and each patch blows off material every time it faces the Sun.

The tails of comets

The tail of a comet begins to appear when the comet approaches the Sun. The average distance of the comet from the Sun at which the tail can be seen is about 1.7 AU and the average length of the tail is of the order 10^7 km. The tail always points away from the Sun; as the comet approaches perihelion the tail trails behind, and after perihelion the tail precedes the nucleus. The tail is being blown away from the Sun by the pressure of solar radiation and of an outward flow of gas – the solar wind. The radiation acts on small particles of dust, and the solar wind on ionised gas, accelerating them to different velocities. Often the dust and the gas form two distinct tails because of these two different velocities away from the Sun. The steady disintegration of the nucleus and coma at every perihelion passage will eventually lead to the disappearance of the comet. As already mentioned the spectral evidence is that the more stable molecules such as carbon monoxide and nitrogen persist in the material blown out into the tail and the rapid motion often observed along the tail is clear evidence of the effect of the solar wind.

The rate of growth of a comet's tail as it moves towards the Sun is extremely rapid. Although the average tail length is quoted above as 10^7 km some cometary tails are much longer. The tail may grow at the rate of 10^6 km per day. The tail of the great comet of 1843 stretched for 3×10^8 km – twice the distance of the Earth from the Sun. In the 1910 return of Halley's comet the tail was 5×10^7 km at perihelion and then continued to increase for several weeks to 1.5×10^8 km.

Comet Bennett photographed by the British amateurs R.L. Waterfield and M.J. Hendrie on the 4th April 1970. Separate dust and gas tails are clearly visible.

The chances of a collision with Earth

No comet contains sufficient mass for it to exert a discernible gravitational pull on the Earth or any other planet. There is nevertheless a small but interesting possibility that there might be an actual collision with the Earth. What would be the consequences of such a collision between the Earth and a body with a mass of, say, some thousands of millions of tonnes moving at tens of kilometres per second?

First, as regards the tail. There is evidence that the Earth actually collided with cometary tails in 1861 and 1910, but there are no records of significant events other than a brightening of the sky. This is scarcely surprising since the density of material in the tail is exceedingly low – in 1000 km^3 of the average tail there is no more material than in 1 cm^3 of ordinary air.

Collision with the nucleus would be another matter. On the average this might occur once every 15 million years. The material of the cometary nucleus is probably so similar to that of a meteorite that it may be difficult to distinguish between the impact of a huge meteorite and that of a cometary nucleus. The Arizona crater with a diameter of 1200 m discovered in 1891 and now a tourist attraction is commonly regarded as the result of a collision with a meteorite of mass 1500 t. More recently a far larger crater was discovered in the Ungava peninsular of eastern Canada. First observed from the air, this cone-shaped crater was reached by scientists in 1950. It rises 168 m above the plain and is over 3 km across. The mass of the body that caused this crater is estimated to have been at least 150 000 t.

The event in Siberia that occurred on 30 June 1908 may have been a cometary impact. Fortunately this was in a remote region, but when it was first reached by scientists in 1927 it was discovered that the forest had been completely devastated over a diameter of 100 km. Houses had been destroyed 250 km from the impact point, the explosion had been heard at distances of over 900 km, and jets of flame could be seen 400 km away. If this impact had

The explosion of a fireball that ripped through the Siberian forest on the 30th June 1908 was heard more than a thousand kilometres away and the effects of the shock wave were felt over a wide area but the first expedition to the remote site was not mounted until 1927 when trees were found to have been felled and stripped of their branches for up to 18 kilometres from the centre of the explosion. The young upright shoots in this photograph had grown in the period since the explosion.

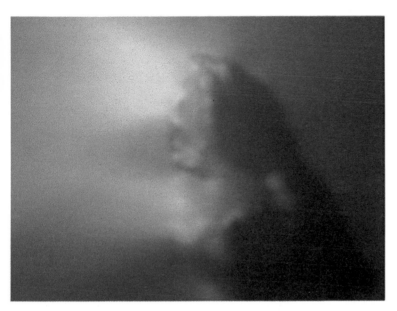

This picture of the nucleus of Comet Halley is composed of seven images taken by the Halley Multicolour Camera during the encounter of the ESA Giotto spacecraft on 13 March 1986 at distances ranging from 25 thousand to 2730 kilometres. The image has been corrected by computer processing and is shown in false colours.

The Giotto spacecraft, which encountered Comet Halley on the 14th March 1986, was launched on the 2nd July 1985 from French Guyana and is seen here mounted on the Ariane 1 launch vehicle.

occurred 4 h 47 m earlier the city of St Petersburg (Leningrad) would have been destroyed. The peculiarity of this event is that no large crater was found in spite of the great extent of damage. The available evidence indicates that the body with which the Earth collided fragmented as it plunged through the atmosphere.

Although the chances of collision with a massive cometary nucleus are so remote, there is little doubt that a world-wide catastrophe could result if the Earth collided with a cometary nucleus of average size. It is possible that such a collision resulted in the extinction of the dinosaurs 65 million years ago. Recent geological evidence obtained from the study of the type and level of soot in ancient sediments suggest that a cometary nucleus about 10 km in diameter collided with the Earth in that epoch. Vast fires were created and the resulting soot blacked out the Sun and cooled the Earth to such an extent that the dinosaurs could not survive.

The greater probability of collision with bodies of smaller size, classed as meteorites, will be discussed in Chapter 11. It is fortunate that such a large fraction of the Earth's surface is covered by ocean, desert or other uninhabited regions so that the chance of a human disaster even in the remote chance of an impact with a meteorite is vanishingly small.

Space probes to the comets

It has long been the desire of astronomers to study a comet at close quarters and this ambition was first realised on 11 September 1985 when an American space probe passed through the tail of Comet Giacobini–Zinner. This probe was launched in August 1978 with the primary purpose of studying the solar wind. It was originally known as ISEE–3 (that is International Sun Earth Explorer). Subsequently it was realised that it would be possible to change the orbit of the probe by making use of the gravitational attraction of the Moon so that it would make a close encounter with the comet. The fly-by of the Moon occurred on 22 December 1984, and on 11 September 1985 the probe, now re-named ICE (International Cometary Explorer), passed through the tail of Comet Giacobini–Zinner, at a distance of 7800 km from the nucleus. The results confirmed the significance of the interaction with the 400 km s^{-1} solar wind. The cometary atmosphere with which the solar wind interacts results from the sublimation of icy surface material of the 1 km diameter nucleus. These molecules would normally have velocities of about 1 km s^{-1} relative to the nucleus. As the gases – mainly water vapour – expand away from the nucleus they undergo complex photo-chemical reactions with sunlight leading to the break up and ionisation of the neutral molecules. The instruments in the ICE showed that the long filamentary nature of the comet's tail resulted from the capture of magnetic flux from the solar wind. An interesting point was that these interacting regions were found at distances of 70 000–100 000 km from the nucleus and that the tail was 25 000 km across.

Five spacecraft passed close to Comet Halley in March 1986, shortly after its perihelion passage. Two were Russian *Vega* spacecraft, which had already made a fly-by of the planet Venus in June

1985 releasing a landing probe and balloons into the planet's atmosphere. On 6 March 1986 *Vega 1* passed 8890 km from the nucleus of the comet, and *Vega 2* passed 8030 km from the nucleus on 9 March. The Japanese probe *Suisei* passed the comet at a distance of 150 000 km on 8 March, and *Sakigaki* at a distance of 7×10^6 km on 11 March. The data from the *Vega* spacecraft enabled the orbit of the European *Giotto* spacecraft to be re-adjusted so that it passed only 600 km from the nucleus of the comet on 14 March. The ICE spacecraft referred to above in connection with Comet Giacobini–Zinner was at its closest to Halley's comet on 25 March at a distance of 28×10^6 km. All the probes passed on the sunward side of the comet.

Remarkable results were obtained from these encounters, particularly from the *Giotto* probe. Fourteen seconds before its closest approach this probe collided with a particle large enough to cause a deviation of 1.8° from the desired attitude. The telecommunication link to Earth was interrupted for 32 min until the on-board dampers reduced the deviation to less than 1° when continuous data was again received on Earth. The spacecraft and some of the on-board equipment suffered some damage during this encounter but it was possible to re-target the probe so that it will return to the neighbourhood of Earth in July 1990. The probe could then be re-directed towards Comet Grigg–Skjellerup for an encounter on 14 July 1992.

Images transmitted to Earth from the multi-colour camera revealed that the nucleus of the comet was non-spherical (described as being shaped like a potato). The major axis was about 15 km and the minor axis 7–10 km. The surface was irregular and of very low albedo, comparable with the darkest bodies in the solar system – presumably because it is covered with a layer of dust. The cometary dust was not emerging uniformly from the nucleus, but from two jets on the sunward side of the nucleus. The jets appeared to emerge from pits about 100 m deep covering about 10% of the surface. The dust is then swept around by the pressure of the solar wind to the tail of the comet. From the successive images of these surface features it was found that the cometary nucleus was rotating with a period of 2.2 days. The neutral mass spectrometer carried in *Giotto* revealed that water was the dominant parent molecule in the comet. Near the time of close encounter the gas production rate was 6.9×10^{29} molecules per second of which 80% were water vapour molecules. Of the other molecules present in the gas, carbon dioxide was prominent at a production rate of about 1.9×10^{28} molecules per second.

The effect of the interaction of the comet with the solar wind (the bow shock) was observed at a distance from the comet of over 10^6 km and magnetic field variations were identified at a distance of 2×10^6 km from the nucleus. The first dust particles from the comet were detected at a distance of 290 000 km from the nucleus and during the encounter 12 000 impacts were recorded of dust particles ranging in mass from 10^{-17} to 1.4×10^{-4} g. The total dust production rate near the closest approach was 3.1×10^6 g s^{-1} (over 30 t s^{-1}). As mentioned on page 110 the implication is that the nucleus of Halley's comet shrinks by about a metre at every perihelion passage.

11

Meteors, micrometeorites and meteorites

Most of us have experienced the thrill of seeing a bright meteor streak across a dark night sky, appearing without warning and fading into nothing before one even has time to point out the trail to a companion. A few fortunate people have seen the brighter trails that actually reach ground level, where on rare occasions a meteorite can be found; these trails are so bright that they deserve the name 'fireball'. These meteors and meteorites are, as we shall see in this chapter, all members of the solar system; they are very important in understanding the solar system, for they provide us with a continuous supply of samples of the primitive material of which the original solar nebula was made. Some authors use the word 'meteoroids' for the solid particles, reserving the word meteor for the visible streak of light also commonly called a 'shooting star'.

Rocks, pinheads and dust

Bright and spectacular though they are, meteors are usually very small bodies, typically about the size of a pinhead. These tiny bodies are, like the planets, in orbit round the Sun. But their orbits, unlike those of the planets, are often far from circular and often highly inclined to the ecliptic plane, in which the Earth and the other planets move. The Earth cuts through many of these orbits, encountering the meteoritic material as it does so. This material

The explosion of a fireball, a member of the Quadrantid meteor shower, captured on film by James Shepherd of Edinburgh on the 4th January 1984. The estimated magnitude was −10.

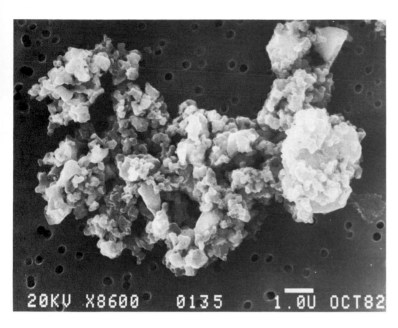

A particle of cosmic dust collected in the stratosphere by a NASA U2 aircraft. The white scale bar measures one micrometre.

covers a very wide range, from particles of less than one-millionth of a gram to large chunks of matter weighing several kilograms. Surprisingly, both the very smallest and the very largest reach the surface of the Earth, while the pinhead sizes are so heated by their headlong rush through the atmosphere that they evaporate at a height of many kilometres. The large metcorites cannot be stopped by the atmosphere, and often arrive intact; the fine dust, in complete contrast, is slowed before it even gets hot, and showers gently down as an almost unnoticed fine rain of primordial particles.

The total mass of meteoritic material of all sizes swept up by the Earth in any year is about 10^6 kg, which is only one-billion-billionth of the mass of the Earth. Two-thirds of this mass falls on the Earth – about equally divided between the large meteorites and the very small micrometeorites. The remaining one-third burns up in the atmosphere and becomes visible as meteor trails. If the solid meteoritic material reaching the Earth's surface could be spread out evenly and undisturbed then it would take about 5000 million years for 2 cm of the material to accumulate.

Observing meteors

Amateur astronomers know well that watching for meteors is usually very tedious, with only a few examples to reward a whole night's work. But they know also that on certain nights of the year meteors are much more frequent, appearing as a 'meteor shower'. This can be a wonderful experience. One of the authors (BL) well remembers such a night. During the evening of 9 October 1946 there were few meteors to be seen – the usual average of a few per hour. Then quite suddenly after midnight the numbers mounted rapidly. The sharp peak in the activity occurred at about 3.40 a.m. when the visual rate approached 200 per minute. By dawn the rate had decreased again to its usual low value. As we shall see, on that

These graphs show the correspondence between the number of radar echoes recorded (lower trace) and the visual (top) and photographic (middle) observations of meteors during the Giacobinid shower on the 10th October 1946.

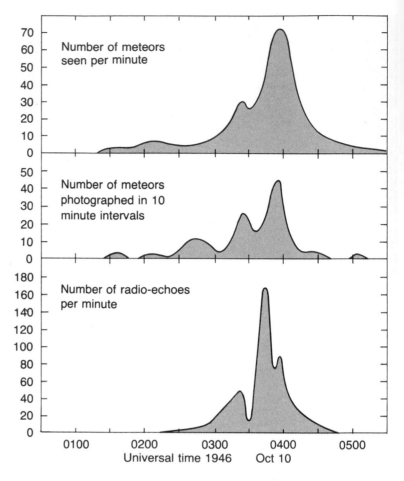

Number of meteors seen per minute

Number of meteors photographed in 10 minute intervals

Number of radio-echoes per minute

Universal time 1946   Oct 10

night the Earth crossed the orbit of a comet, close to the comet itself.

Recording meteors is a demanding but very useful task for the amateur. Special skills must be developed, and serious work is best carried out in collaboration with an amateur organisation such as the British Astronomical Association or your local astronomy club. An individual observer should undertake to watch a defined area of sky, noting the positions of the start and end of each trail against the background of the stars. The brightness of the trail can be estimated by comparison with the magnitude of nearby stars. Of course it is essential to have a detailed knowledge of the names, positions and magnitudes of the stars in the area of the sky under observation. It is also possible to estimate the velocity with which the meteor enters the atmosphere by measuring the time from the beginning to the end of the trail. Since this interval is rarely more than a fraction of a second this method of measuring meteor velocities is not very accurate. Amateur observers should also record the duration for which the trail lasts. Again this is less than 1 s for most meteors, but occasionally much longer lasting trails are seen; rarely, drift trails can be seen for many minutes. These are associated with meteors that are so massive that they reach the lower atmosphere before burning up completely.

This trail of a Geminid meteor, estimated to be magnitude −1, was photographed by Steve Evans of Wiltshire, England, during a one-hour exposure on the 11th December 1986. The breaks in the meteor trail are created by a rotating shutter in front of the lens and enable the speed of the meteor to be calculated. The unbroken trails are those of stars.

Photography can be used, exposing the film in a wide-angle lens camera for as long as possible, before the film becomes completely fogged by background light from the sky. However, unless specially constructed wide-angle lens cameras are used with film of high sensitivity the yield of photographed trails will be low compared with those observable with the unaided eye. Nevertheless, photography provides a good method of measuring velocities if a rotating shutter is used in front of the lens so that the photograph of the trail is occulted several times per second.

The rates of arrival of both sporadic and shower meteors are greater in the second half of the night. This is a result of the combination of the Earth's rotational speed with its orbital speed; in the early part of the night the observer faces away from the direction in which the Earth is moving in its orbit, and fewer meteors are swept up as the Earth cuts through the meteor orbits. A useful analogy is that the windscreen of a moving car collects more raindrops than the rear window. It happens also that the sporadic meteor rate is higher in the second half of the year. This rate depends on the inclination of the Earth's axis to its motion in its orbit which varies through the year.

Meteor showers

The track of a single meteor tells only part of the story of its orbit round the Sun. In a meteor shower, however, we are seeing many

Meteor trails seen among the parallel
star trails in a time exposure. The trails
are at various angles but they all diverge
from a single point, known as the
radiant. This photograph of the Leonid
shower of November 1966 was made by
D. Milon at Kitt Peak, Arizona.

meteors which follow almost the same orbit, but we see them entering the atmosphere at different points all round us. A wide-angle
photograph of many meteors in a single shower shows immediately
that they all seem to come from a single point in the sky, known as
the radiant.

The radiant for each shower is fixed with respect to the stellar
background, so that meteor showers are named after the constellation in which the radiant lies. The precise position of the radiant can
be specified, as for a star, by the coordinates of right ascension and
declination. Furthermore, the meteors in a shower are not only
moving in parallel paths: they all move with the same velocity. The
conclusion is that a meteor shower occurs when the Earth passes
through a swarm of particles all moving in an elliptical orbit around
the Sun, which, of course, differs for the various showers.

A knowledge of the co-ordinates of the radiant R and of the
velocities of the meteors enables the parameters of the orbit of the
particles around the sun to be computed. As will be seen, many of
these orbits are clearly identified with the orbits of certain elliptical
comets. Table 11.1 gives a list of the showers that a normal
observer can expect to see with the unaided eye under clear, dark
sky conditions (no moon). A more comprehensive list providing
estimates of the rate expected and other details is published
annually in the *Handbook of the British Astronomical Association.*
Most of the showers in this list appear annually with about the
same hourly rate. The Leonids are a notable exception. There were
very rich occurrences in 1799, 1833, 1866 and 1932. The peak

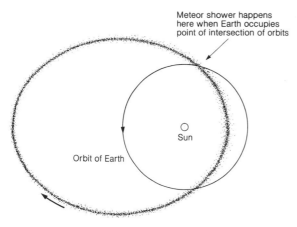

Meteor shower happens here when Earth occupies point of intersection of orbits

Sun

Orbit of Earth

Meteor showers occur when the Earth on its path around the Sun intersects the orbit of a meteor stream.

intensity is reached in 33.2 year periods indicating that, unlike the Perseids or Geminids for example, the debris is not uniformly distributed around the orbital path of the stream. In addition to these regular showers there are a few cases of remarkable periodic meteor showers which will be referred to later in connection with their cometary association.

Table 11.1 *Easily observable meteor showers*

Name of shower	Date of maximum	Radiant co-ordinates rt. ascen.	declination	Geocentric velocity (km/s)
		h min		
Quadrantids	Jan. 3/4	15 28 (232°)	+52°	39
Lyrids	Apr. 21/22	18 08 (272°)	+33°	51
η-Aquarids	May 6	22 20 (335°)	+ 1°	66
δ-Aquarids	July 28	22 36 (339°)	−15°	40
Perseids	Aug. 10–14	03 04 (046°)	+58°	61
Orionids	Oct. 20–23	06 24 (096°)	+15°	66
Taurids	Nov. 3–10	03 44 (056°)	+15°	27
Leonids	Nov. 16–17	10 08 (152°)	+22°	72
Geminids	Dec. 13–14	07 28 (112°)	+32°	35
Ursids	Dec. 22/23	14 28 (217°)	+77°	38

Radar observations of meteors

Until the end of World War II there was an almost total reliance on amateurs for the observation and collection of meteor data. Then it was found that radar observations could be made of the ionised trail left by the meteors as they burnt up in the high atmosphere. The story of this discovery is a remarkable illustration of the manner in which unexpected scientific developments often occur. Before the war American scientists using radio methods to study the ionised regions which surround the Earth at altitudes of 80 km and upwards had noticed occasional increases in ionisation when a bright meteor appeared. However, the relation was inconclusive and doubts were expressed that the observed effects had any connection with meteors.

This entire subject was revolutionised by a remarkable series of events in World War II. The Germans began their bombardment of England with the V2 ballistic rockets on 9 September 1944. These ten tonne rockets carrying a tonne of explosive reached an altitude of nearly 100 km in a ballistic trajectory and five minutes after launching dropped on their target at a speed of 5600 km h^{-1}. No defence was possible but it was realised that if the rocket could be detected in flight then sufficient warning could be given to save many civilian casualties. With the help of the Army Operational Research Group radar sets working on a wavelength of 4.2 m, normally used to detect enemy aircraft so that anti-aircraft guns could be directed against them, were hastily converted for the detection of the V2 rockets. These converted radars could indeed detect the rocket in flight as a short-lived echo on the cathode ray tube but it was found that many transient echoes were seen and warnings of an attack given when no V2s arrived. J.S. Hey, then a member of the AORG staff, investigated this peculiarity and concluded that the transient echoes observed in the absence of V2s were probably similar to the transient ionospheric effects found by the pre-war American scientists.

After the end of World War II in 1945 Hey and Stewart modified the standard army radar equipment to begin a systematic study of these transient echoes and soon concluded that meteor trails were at least a significant cause. Similar radar equipment was set up at Jodrell Bank towards the end of 1945. Although intended for quite different researches, this equipment soon placed the issue of the meteor–transient echo associations beyond doubt. During the Perseid meteor shower of August 1946, J.P.M. Prentice, a solicitor in Stowmarket, Suffolk, and then the director of the meteor section of the British Astronomical Association, carried out visual observations of the meteors outside the trailers of radar equipment. A direct association between visible meteors and transient radar echoes of duration 0.5 s or greater was found but there seemed to be no similar direct connection between the much larger number of shorter duration echoes and the visible meteors. Two months later the issue was settled in a dramatic way. On the night of 9–10 October 1946 Prentice concluded that the Earth would cross the orbit of the Giacobini–Zinner Comet only 15 days after the comet had passed the intersection of its orbit with that of the Earth. This comet was first observed in 1900 and Prentice himself had discovered a strong meteor shower associated with the cometary orbit in 1933. Since the period of the comet is 6.6 years it was already apparent that the meteoric debris was not distributed along the orbit but must be still concentrated in the region of the comet itself – thus Prentice expected a very active shower to occur in October 1946.

The radar results on that night of 9–10 October were extraordinary. Until midnight only the normal low sporadic meteor echo rate of a few per hour were observed. Suddenly the echo rate began to increase rapidly so that at the peak activity at 3.40 a.m. on 10 October the rate was about 200 per *minute*. By 6.30 a.m. the activity had once more decreased to the normal sporadic background rate. Outside the sky was vivid with meteor trails and near the peak the narrow-beam antenna was directed at the posi-

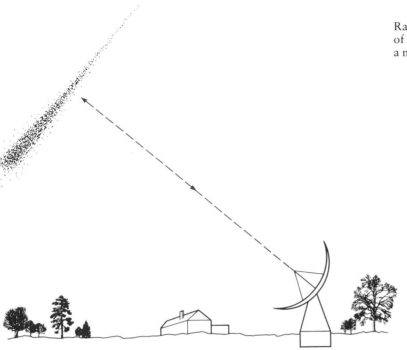

Radar echoes are reflected from the trail of ionized gas created by the passage of a meteor through the atmosphere.

tion of the shower radiant. In this position the echo rate decreased abruptly but was immediately restored when the antenna was returned to its previous position where the meteor trails were forming at right angles to the direction of the aerial beam. The observations during that night clarified two important issues: first, that all the transient echoes were associated with the meteor shower, the implication being that the radar was sensitive to trails below the naked eye visibility limit and, second, that the ionised trail of the meteor reflected the radio waves by broadside, or specular, reflection.

There were very important consequences of the recognition that the transient echoes were clearly associated with meteor trails and that the sensitivities of the radar enabled ionised trails to be detected from meteors that were too faint to be observed visually. The visual and photographic techniques were useful only under clear sky conditions and were also severely handicapped in the half to full moon period. Thus the accumulation of reliable systematic data about meteor showers was exceedingly difficult. The radar observations were not affected by cloud or moon and, of even greater significance, they could be used in the daytime. During the summer of 1947 a series of most active daytime meteor streams were discovered at Jodrell Bank. The radiants of these showers did not rise before dawn and hence no previous knowledge of these showers could have been obtained by visual or photographic observations. The specular reflection properties of the trails enabled radar methods to be developed for the location of the radiant point, and perhaps of greatest importance was the development of a method for the precise measurement of the velocities of individual meteors.

Although radar measurements of meteors are beyond the scope

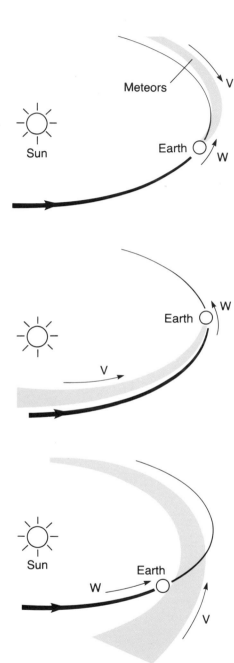

The velocity with which a meteor enters the atmosphere is a combination of its own velocity and that of the Earth so its speed will depend on the direction from which it comes. A fuller explanation is given in the main text.

of amateur observers, the effect of the formation of the ionised trail can be readily detected as a 'whistler' if the observer is within the ground range of a short wave transmitter. This is a continuous wave (cw) as distinct from a pulsed radar effect and is most easily observed from cw transmitters in the 10–20 MHz band. Provided the strength of the ground wave from the transmitter is not too great then an interference beat can be observed from this ground wave and the radio wave reflected from the ionised trail of the meteor. The phenomenon can be heard in headphones or a loudspeaker as a 'whistle' decreasing rapidly from a few hundred to zero hertz. When these were first heard they were believed to be a Doppler whistle caused by the reflection of the cw signal from the ionisation at the head of the moving meteor. However, it is now recognised that this is not the correct explanation; the whistle arises from the amplitude and frequency changes in the diffraction pattern created as the ionised column of the meteor trail develops. Sophisticated developments of this cw technique have been used as well as the pulse method to make accurate measurements of the meteor velocities.

The velocity of meteors

The shower meteors are clearly all in elliptical orbits round the Sun. The sporadic meteors cannot so easily be assigned to such orbits, and it was for some time thought that many of them might come from outside the solar system. The question could only be settled by reliable measurements of their velocities, and the necessary accuracy was finally achieved by radar methods.

The velocity of a meteor entering the Earth's atmosphere is a combination of its own orbital velocity and the velocity of the Earth. Only the first component, that is the orbital velocity of the meteor around the Sun (known as the heliocentric velocity), need be established to decide whether or not the meteor is moving in a closed orbit around the Sun. However, in all the actual observations of meteor velocities the Earth's orbital velocity of 30 km s^{-1} is combined with this heliocentric velocity of the meteor. The *measured* velocity of the meteor therefore depends on the direction with which it enters the atmosphere relative to the direction of the Earth's motion in its orbit. This measured velocity is known as the geocentric velocity of the meteor, and, as can be seen from the diagrams, this can vary within wide limits.

In the first case at (*a*) the orbit of the meteor is such that it will make a head-on encounter with the Earth. The measured geocentric velocity as it enters the atmosphere will then be the sum of the Earth's orbital velocity and that of the meteor around the Sun, that is $(W + V)$. In the second case at (*b*) the meteor is following the Earth in its orbit and the measured geocentric velocity will be $(V - W)$. Of course, in practice, most of the cases observed will be intermediate between (*a*) and (*b*). There is a critical value, 42 km s^{-1}, for the heliocentric velocity (V) of a meteor. If the velocity of the meteor exceeds this value then it cannot be moving in a closed (elliptical) orbit around the Sun and must have originated from outside the solar system. In practice this means that all meteors that are permanent members of the solar system must have observed (geocentric) velocities lying between the extreme cases of (*a*) (30 + 42) = 72 km s^{-1} or (*b*) (42–30) = 12 km s^{-1}. If any meteors were found to have geocentric velocities greater than 72 km s^{-1} then their heliocentric velocities must be greater than the escape velocity of 42 km s^{-1}. Radar observations of many thousands of sporadic meteors excluded all reasonable possibility that this could be the case and we now regard all sporadic meteors as part of the solar system.

Ironically the radar measurements showed that many of the sporadic meteors were very far from being extra-solar; the very faint ones especially are moving in unusually short period orbits. Many are in nearly circular orbits which are concentrated near an inclination of 60° to the plane of the ecliptic. One can only speculate on the close encounter with a planet that might have thrown them into this unusual situation.

The observation of micrometeorites

It has already been mentioned that about one-third of the total mass of meteoritic material swept up by the Earth is in the form of micrometeorites. These are meteor particles that are so small that their energy is dissipated before they burn up in the atmosphere and they then fall to Earth in the form of minute grains of dust.

Although this fall-out is quite invisible it is a fairly simple matter to establish the existence of these microscopic particles. During the course of a week or so the sediment deposited by the air can be collected in a suitable container. Near industrial or urban areas this deposit will consist largely of particles originating from Earth. On the other hand in districts far away from habitation the sediment will contain a significant quantity of the micrometeorites from space. These micrometeorites contain ferro-magnetic material and can be separated from the remaining sediment by using a magnet. They are generally very small (less than 120 μm in diameter) but can be studied by using a microscope. If this simple experiment is carried out after one of the major meteor showers then, allowing for the time taken for these particles to fall through the atmosphere, the number of ferrous particles in the sediment will show a marked increase.

Considerable information has been obtained about these micro-meteorites by analysis of the sediments from the polar ice caps and from the ocean beds. When collecting from ocean sites it is possible to choose locations remote from volcanic and other terrestrial fall out. In one recent investigation of the floor of the ocean the particles were collected by dragging a 300 kg magnet across the sea floor at a depth of 5 km.

The particles collected from fall out on the Earth may either be the unaltered particles from space that are small enough to have been stopped in the atmosphere before melting or larger particles that have survived after partial melting during their descent. A much clearer picture of these primeval microscopic dust particles is emerging now that it has become possible to carry out the investigations in earth satellites and space probes. For example in the American *Skylab* flight thin foils and polished metal plates were exposed in space and when returned to Earth the impacts were studied with optical and electron microscopes. The analysis of the cratering on these foils produces data on the number, size and velocities of the micrometeorites. In cases where the spacecraft does not return to Earth impact detectors have been employed. Modern forms of impact detectors make use of the plasma generated by the particles when they collide with the target. A form of mass spectrometer is then used to determine the mass of the particle.

Meteorites

As we have seen the microscopic particles forming the micro-meteorites reach Earth because their impact with the atmosphere does not generate sufficient energy to cause complete evaporation, or in extreme cases their flight is halted before any melting occurs. At the other extreme range of sizes we have meteorites – lumps or fragments of mineral and metallic composition which are far too massive to be evaporated in their passage through the Earth's atmosphere. It is estimated that every year about 100 fragments of meteorites fall on the United States from bodies originally weighing more than 5 kg before entering the atmosphere. A body of this size would first be seen as an extremely bright meteor (or fireball) penetrating to the lower atmosphere where the material stresses become so great that it may well fragment into many smaller pieces which then reach the ground. More rarely the Earth encounters much more massive objects. It is estimated that once in 30 years the Earth may collide with bodies of 50 tonnes or more, several metres in size – about one every 500 years falling on an area as large as the United States. Fortunately of even greater rarity are the encounters with bodies which have masses of several hundred tonnes. About one such body of this size plunges to Earth every 150 years or so – or once every 2500 years on a land mass the size of the United States.

Meteorite falls of this mass or larger could be cataclysmic events if the body came to Earth near an inhabited centre. Indeed there is ample evidence of the devastation that could be caused by these gigantic encounters. The famous meteor crater in the Arizona desert is a huge pit nearly 1200 m across and 180 m deep. It is estimated that the meteorite made this impact many thousands of years ago. Of more recent occurrence was the tremendous impact of 30 June 1908 in Siberia. As already explained in Chapter 10 this may have been a collision with the nucleus of a comet as distinct from a typical large meteorite.

Because of the information meteorites give us about the constitution of solar system bodies their recovery and preservation for expert analysis is a matter of great importance. Anyone fortunate enough to witness a fireball of exceptional brilliance in the sky should make a careful note of its trajectory and the apparent place of cessation of the trail in the sky. With such information, particularly if records are available from different places, it is frequently possible to locate the area of impact with the Earth with sufficient accuracy so that a search leads to the recovery of fragments of the meteorites. Any meteoric material recovered should be immediately sent for professional examination (for example in the UK to the Natural History Museum) since, apart from the relevance to the

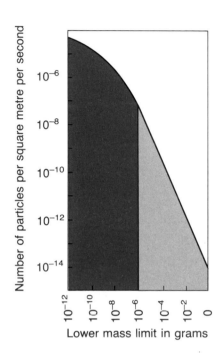

This graph shows the distribution in size of the meteors in the Earth's vicinity in space. The cumulative total, that is the total number of meteors in excess of the mass shown along the horizontal axis, is plotted in the vertical direction. Both scales are logarithmic. The graph shows that there are far more microscopic meteors than ones of sizes perceptible to the eye.

investigation of the primeval condition of bodies in the solar system, there is at the moment, as we shall see later, a great interest in the search for extraterrestrial organisms in meteorites that have fallen to Earth.

In general it is found that about 61% of the meteorites are stony, about 35% are iron and the remaining 4% are of a stony–iron composition. Many of the stony meteorites are often referred to as chondrites because they contain small spherical inclusions (chondrules), indicative of formation during rapid cooling. Occasionally chondrites are black and contain a few per cent of carbon. These are the carbonaceous chondrites.

The famous crater in Arizona is one of the youngest impact craters on the Earth. The bowl-shaped crater, about 1.2 kilometres across and 200 metres deep was formed about 25 000 years ago. The iron meteorite fragmented and was dispersed in the impact.

Origins

The combination of the various visual, photographic, radar and space methods of investigation enables fairly reliable estimates to be made of the numbers of meteors of the various sizes which exist in space and which the Earth sweeps up in its orbit around the Sun. For example, even excluding the micrometeorites, it can be estimated that about 8×10^6 meteors enter the Earth's atmosphere per day – that is of sizes that can be detected by radar or be seen visually.

Although these numbers are large, the actual concentration of the particles in space is very low. From the graph it can be calculated that a particle of mass 1 mg occurs on the average only once in 10^7 mile^3 (10^{23} cm^3). A particle of mass 1 g is likely to be encountered only once every 10^{10} mile^3 of space (10^{26} cm^3). Thus the danger of serious impacts with spacecraft is extremely small.

This meteorite was found in 1931 by
H.H. Nininger in Saskatchewan,
Canada. It is thought to be one of a
hundred or more pieces of which a
dozen were recovered. Rounded crystals
of olivine can be seen in this stony
meteorite.

Although it is now well established that these particles of all sizes
are confined to the solar system the question of how they originated
is a most important one. There is now a belief that the massive
meteorites that fall to Earth are associated with the asteroids.
In the case of the shower meteors, the evidence for a close association with comets is quite clear. In many cases the orbits of the
shower meteors are the same as those of known comets. A recent
discovery of an asteroid by the infrared satellite IRAS provides
another example: it appears to be in the same orbit as the Geminid
meteor shower.

The orbits of the sporadic meteors being of short period are also
similar to the orbits of the asteroids, but the origin of this large
amount of dust remains uncertain. It is not yet known whether this
sporadic meteor dust is a primeval remnant or the result of a subsequent planetary or cometary break up. An argument against a
primeval origin is that no particle with radius less than about 4 cm
can have survived in the solar system since its origin some 4.5
billion years ago, because of the effect of sunlight on its orbit. The
existing particles cannot therefore have the same primeval origin as
the solar system. This problem is particularly acute for meteoritic
debris moving in orbits of short period.

Although this particular problem is unsolved, it seems to be a safe
assumption that meteors, micrometeorites and meteorites of all
sizes are associated either with comets or asteroids.

12

The Sun

Life depends on the heat and light of the Sun, but we hardly ever look at it directly. Indeed it is dangerous to do so, with or without a telescope, except in unusual conditions of a hazy sky near sunset. Doubtless the ancient Chinese observers who first reported sunspots were seeing the Sun through the atmosphere very near the horizon; until the time of Galileo it was thought that these spots were actually in the Earth's atmosphere, not on the surface of the Sun. The Sun is nevertheless safe to observe by projecting the image on to a white surface, using a small telescope or even one half of simple binoculars. (Try a distance of 0.5 m, with the telescope focussed somewhat short of infinity. A crude mounting to steady the telescope is a great help.)

Large solar telescopes may project an image up to 1 m across,

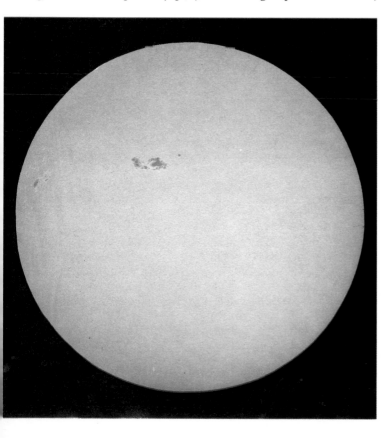

The disc of the Sun looks darker around the edge (what astronomers call the 'limb') because the Sun is not solid but a ball of gas. A large sunspot group can also be seen.

and portions of this image may then be examined in a spectrograph. The enormous amount of light available from the Sun allows the spectrum to be examined in very great detail. Our simple projected image nevertheless can tell us a great deal about the Sun. Look, for example, at the edge of the bright disc: it is appreciably darker than the centre. Compare the colour and the brightness with an electric arc (but be very careful not to look directly at it). The colour and intensity of the light reveal the temperature of its source: the Sun's visible disc (photosphere) is at about 5800 K while the arc is cooler even than the sunspots, which are around 4000 K. The edge of the solar disc appears darker than the centre because the radiation is coming from higher and cooler layers.

This cooler region is the chromosphere; without it the Sun would be uniformly bright, and the spectrum of its light would be relatively featureless. A spectrograph shows, however, that the spectrum contains an almost continuous series of absorption lines, corresponding to many different elements. About two-thirds of the known elements are represented. Most of these elements are not at all abundant, but they happen to be efficient in absorbing the light from the photosphere below. We will later be discussing the origin of these elements; they certainly are not synthesised from the hydrogen and helium of the Sun, and it is reasonable to suppose that they are representative of the original nebula from which the Sun was formed.

There is a steady drop in the temperature of the chromosphere with increasing height above the photosphere until, at a distance of about 2000 km, the temperature rises rapidly to over 10^6 K. This very hot and very tenuous region, called the corona, extends to a distance of several solar radii. It can be seen with the naked eye at a solar eclipse, which is a rare phenomenon well worth making an effort to see. The corona is as bright as the full moon, but usually the light from the Sun's surface is too bright for it to be seen.

The Sun also emits radio waves. A fairly simple radio telescope can be used to measure the intensity of solar radio emission; even a sensitive television receiver may sometimes detect solar radio noise, especially at the time when large sunspots can be seen. On short radio wavelengths, around 1 cm, the radio emission is from near the surface. On long wavelengths, around 1 m, the emission is from the hot corona, and the intensity is correspondingly higher. Furthermore, the corona is so much larger than the photosphere that the radio Sun is a large, diffuse object more than 3° across, as compared with 0.5° for the photosphere.

Not only radio waves but also electromagnetic waves from practically the whole observable spectrum are emitted by the Sun. X-rays, which can only be observed from spacecraft, are particularly informative. Like long radio waves, they originate in the corona; unlike radio waves, they include spectral lines from several different elements. The corona is so hot that these elements are very highly ionised; for example, oxygen atoms may be stripped of six of their total of eight electrons, and iron atoms may lose 15 out of 26 electrons. X-rays offer our best view of the inner corona, showing it to be very patchy on a large scale. In particular, there are very large dark patches, known as coronal holes.

Solar energy

One can get equally hot by sunbathing or by standing in front of an electric fire, and this comparison suggests that the flux of solar energy falling on the Earth is about 1 kW m^{-2}. To find the total power emitted by the Sun one can either use that figure and the radius of the Earth's orbit, or one can calculate the so-called 'black-body' radiation from a sphere at a temperature of 5800 K. We then find that the Sun radiates more energy in one second than mankind has consumed in the whole of our history. This energy comes from a nuclear furnace at the centre of the Sun, where 6×10^8 tonnes of hydrogen are transformed into helium every second. About 1% of this mass is lost in the transformation and appears as the radiant energy on which life on Earth depends.

This seemingly enormous rate of hydrogen burning is fortunately a small rate of loss for the Sun, which can continue at the same rate for 5000 million years. The mass of the Sun is 330 000 times that of the Earth, and about three-quarters of it is still unburnt hydrogen. Most of the rest is helium, and most of that helium probably existed before the solar furnace started to burn. The Sun is still young, and it has a long life before it.

Hydrogen burning requires a temperature of about 1.5×10^7 K. These nuclear reactions occur in the core of the Sun, and it is possible to obtain an idea of the internal structure of the Sun from the observation that the observed surface temperature of the Sun is 5800 K. This fall in temperature as we progress outwards from the core is determined by a balance between gravity and pressure, and by the outward progress of the radiant energy. From this it is deduced that the transport of energy from the interior is by radiation for about five-sevenths of the radius. The remaining two-sevenths (about 200 000 km) is believed to be a convective zone. The boundary between the radiative and convective zones is probably ill-defined and may be the seat of an intense magnetic

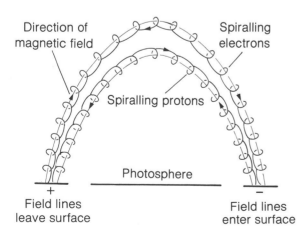

Loops of strong magnetic field break the surface of the Sun at places that become recognizable as sunspots. Charged particles spiral around the field lines.

The only really safe way to observe the Sun visually is to project its image onto a screen, shaded from the direct sunlight.

The internal structure of the Sun. Energy generated by nuclear reactions in the interior works its way outwards through a zone where radiation is the main mechanism to a turbulent convection zone.

The solar corona, becomes visible only when the glare of the Sun's disc is blotted out during a total eclipse. This one took place on the 31st July 1981 and was observed from the USSR. Prominences can also be seen leaping outwards from the Sun. It is a pure fluke of nature that the Sun and Moon have just about the same apparent angular diameter in our sky, making the remarkable phenomenon of a total eclipse possible.

The hot solar corona is an intense source of X-rays. This colour-coded X-ray image of the Sun was obtained by the United States orbiting Skylab.

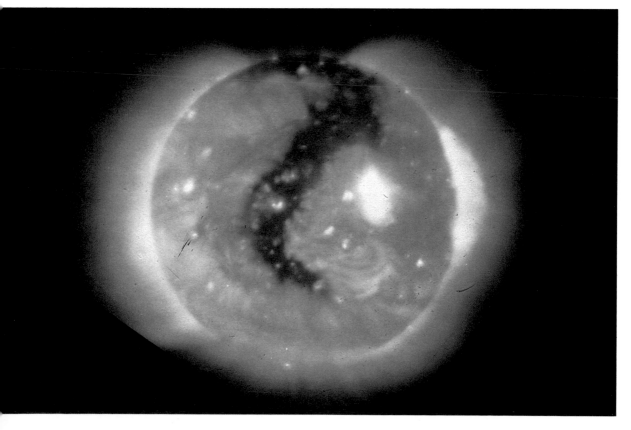

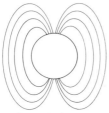

A section through the magnetic field of the Sun shows that it broadly resembles that of a bar magnet.

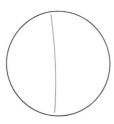

To look at the effect of the Sun's rotation on the magnetic field, we first single out one typical field line.

This line is carried round with the ionized gas in the Sun's upper layers as it rotates.

Because the Sun rotates fastest near the equator, the line becomes bent.

After many rotations, the line is wrapped tightly at low latitudes, resulting in an east–west field.

field. The hot furnace generates mainly gamma rays, which could destroy life on Earth if they were not absorbed and re-radiated with lower energies as the solar energy progresses outward. It takes hundreds of thousands of years for radiation to reach the surface.

The burning process is understood very well, except for one discordant observation. One of the products of hydrogen burning is an enormous flux of neutrinos, which are almost undetectable elementary particles. They travel straight out from the centre of the Sun without interacting with the rest of the solar material. Neutrinos similarly pass straight through the Earth, and are seemingly impossible to detect despite their very large numbers. There is, however, a very rare reaction between neutrinos and chlorine atoms, in which the nucleus of the chlorine isotope ^{37}Cl is transformed into ^{37}Ar, i.e. into an argon atom. This isotope of argon is radioactive, and its decay can be detected.

A remarkable experiment to detect solar neutrinos by their reaction with chlorine has been conducted for several years by Raymond Davis in a deep mine in South Dakota, 1500 m below ground. A large tank of ethylene perchlorate is kept away from all spurious sources of radiation by the overlying rock, and only neutrinos can penetrate that far. Every week the tank is flushed through to collect argon gas. This gas is investigated for individual atoms undergoing the typical radioactive decay. The result is that a few neutrinos can be detected every week. To the consternation of the theorists, the result is a factor of two smaller than expected. A good explanation is still awaited.

Sunspots and rotation

So far we have treated the Sun as a symmetrical, uniform sphere. The projected image from our simple telescope will easily show that this view ignores a remarkable irregularity: the sunspots. These dark areas typically cover a fraction of 1% of the solar surface; often they occur in groups of up to ten individual spots. A daily sketch of their positions shows that the Sun is rotating; sunspots cross the disc in about 25 days if they are at the higher solar latitudes. Allowing for the motion of the Earth round the Sun we find that the Sun is rotating in 26 days at the equator and in 31 days at latitude 60°.

This differential rotation must extend deep into the Sun, since each spot is an area of very strong magnetic field which couples it to the interior. There is a general magnetic field throughout the Sun, and the differential rotation amplifies the general field by a screwing action. A loop of amplified field breaks the surface, producing a pair of spots where the field lines leave and re-enter the surface.

A longer series of observations will show a systematic variation in the number of spots, and the latitude in which they appear. The variation is cyclic, with a period of 11 years. Some features alternate between cycles, so that the complete sunspot cycle lasts 22 years; this is probably the period of a massive oscillation of the magnetic field in the interior.

The spots mainly occur in latitudes between 55° and 25°, both north and south. At the start of a cycle they tend to occur at higher

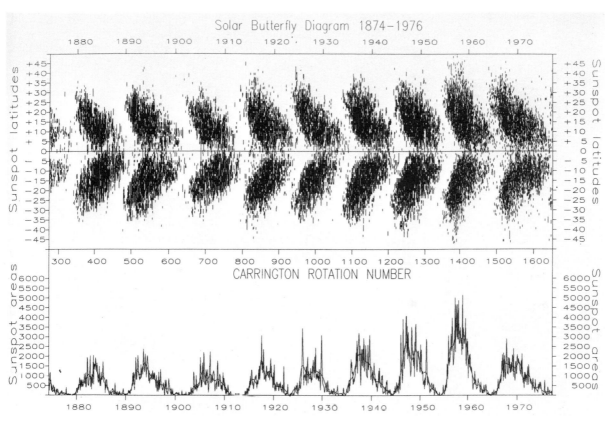

Solar Butterfly Diagram 1874–1976

latitudes, working down to lower latitudes through the cycle. For obvious reasons, the diagram illustrating this is referred to as the 'butterfly diagram', even though it is officially called the 'Maunder diagram' after its originator, Edward Walter Maunder, who was Superintendent of the solar department of the Royal Observatory, Greenwich, from 1873 to 1913.

The active Sun

Spectroscopic observations show that all sunspots have large magnetic fields, typically reaching 3000 gauss. These fields are changing and moving; they must therefore generate large electric fields. The resultant activity can be observed in many ways, particularly when the electric fields create a massive arc discharge in the sunspot regions. This appears as a solar flare, lasting a few minutes or, occasionally, some hours.

On very rare occasions a solar flare can be seen on an ordinary white light projected image of the Sun. The light from a flare is, however, concentrated in a few spectral lines, and it becomes much more obvious if the Sun can be observed only in, say, the Hα line of hydrogen. Huge explosions are then seen, usually stretched along great arcs following magnetic field lines above the sunspots. When these flares occur near the limb of the Sun they can be seen to extend high into the corona, moving outwards at great speed. The arch prominence of June 1946 reached a height of 400 000 km above the surface in half an hour.

If the solar latitude of each sunspot is plotted against time, the result is the so-called 'butterfly diagram', which illustrates how, during the course of the 11-year solar cycle, the number of sunspots and latitudes at which they are observed vary in a systematic way. This plot from the Royal Greenwich Observatory covers the years 1874 to 1976 and also includes a graph of sunspot areas.

Solar flares also generate X-rays and intense radio emission. The radio emission at metre wavelengths can interfere with radio communications and radar. The initial discovery of radio flare emission was associated with a remarkable incident in World War II. On 12 February 1942 the German battleships *Scharnhorst* and *Gneisenau* succeeded in moving through the English Channel from Brest to their home ports in Germany without coming under attack. They had not been detected by the chain of coastal defence radars along the south coast of England. Investigation revealed that the radars had been jammed and the problem was referred to the scientists in the Army Operational Research Group (AORG). Two weeks later J.S. Hey, who was in charge of this investigation, received reports of apparently similar jamming of some anti-aircraft gun laying radars working on a wavelength of 4.2 m. However, the anticipated German air attacks did not materialise and there seemed to be no reason why the Germans should have jammed these radars. Hey then noticed that the reports of this jamming came only from a certain number of radars, and in searching for a common cause he discovered that the directional aerial arrays of the radars reporting the jamming were all pointing towards the Sun. He consulted the authorities on the Sun at the Royal Greenwich Observatory and was informed that on 28 February a large sunspot group had been on the central meridian of the Sun and that there had been an associated solar eruption – a solar flare. He concluded that the apparent jamming of the radars must have been caused by intense radio emission from the flare.

During a solar flare the peak of the radio emission moves from high to low radio frequencies, corresponding to the resonant frequency of successively higher parts of the corona. Flares can be traced by this radio emission out to several solar radii. There is no reason to think that the activity stops there: in fact we know that the Sun is the origin of continuous and variable streams of particles, only some of which come from flare activity.

The solar wind

Every second about 10^6 tonnes of hydrogen are blown away from the Sun. By the time this solar wind reaches the Earth it is very diffuse, with only about 10 particles per cubic centimetre, but it is moving at about 400 km/s. Although it is so diffuse its effects are dramatic and easily observed.

The first suggestion that electrically charged particles were reaching the Earth from the Sun was made in 1896 by K. Birkeland, a Norwegian scientist who was concerned to explain the aurorae. There is a simple relation between aurorae and sunspots: aurorae follow the sunspot cycle, and particularly good displays are seen several days after a major solar flare. A similar relation was already

---

Opposite A solar flare occurs when energy stored in the magnetic field associated with an active area of the Sun is suddenly released in an explosive event in the solar corona. The bursts of electrically charged particles from flares disturb the Earth's magnetic field and cause increased auroral displays.

known between flares and disturbances to the Earth's magnetic field, and another such relation was found later between flares and disturbances to radio communications.

The most obvious effect of the solar wind can be seen in any bright comet when it is near the Sun. As described in Chapter 10, the comet tail always points away from the Sun: it is literally blown out in this direction by the solar wind.

The origin of the steady wind is easily understood as evaporation from the very hot solar corona. The outflow is, however, very varied across the Sun. This is due to a complex magnetic field structure, which interacts with the ionised gas and channels it into streamers. Photographs of the corona extending out to 11 solar radii, using a coronagraph camera in a high flying aircraft, show these streamers very clearly. At sunspot minimum they are concentrated in the equatorial region, but at sunspot maximum they are seen at all latitudes.

Looking directly at the surface of the Sun with an X-ray camera, the streamers are seen to originate in hot spots which persist for many solar revolutions. The structure of the streamers, and of the magnetic field carried with them, can be detected in interplanetary space by direct measurement from space probes. The solar wind has been detected in this way to the distance of the planet Saturn. The streamers of ionised gas also affect the passage of radio waves from distant objects such as the quasars, causing a scintillation analogous to the familiar twinkling of the visible stars.

Our nearest star

Seen from a distant point in the Milky Way, the Sun would be an unremarkable and ordinary star. Technically it would be classed as a dwarf G type, inconspicuous compared with nearby giants such as Aldebaran and Betelgeuse. Well established theory treats the Sun as a normal star, whose age is known and whose evolution is predictable. We are, however, situated so close to this rather ordinary star that we can observe it in very great detail, and we find many aspects of it that are hard to understand.

A major problem is the temperature distribution above the photosphere. How can the corona be so hot, when it is separated from the photosphere by a cooler layer, the chromosphere? A suggestion that very energetic sound waves cross the chromosphere and dissipate their energy in the corona seems an inadequate explanation. It is now believed that a wave motion involving the magnetic field must be responsible, but this cannot be understood without further investigation of the structure of the magnetic field at the surface. This structure can be deduced from the fine details of the visible surface, such as the granulations which cover the whole surface. Solar telescopes with very good angular resolution are needed to observe the granulation; despite the advent of space vehicles to carry telescopes above the atmosphere, the best chance seems to be to use ground-based telescopes on high mountain sites.

A recent discovery of oscillations on the solar surface gives a new possibility of exploring the interior of the Sun. These oscillations are seen in the photosphere, but their very long periods, including a

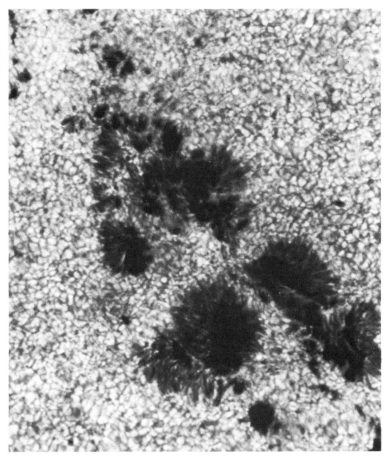

The disc of the Sun is not uniformly bright between sunspots but has an ever-changing granular appearance. The granules, which are typically 1500 kilometres in size, look brighter because they are about 400 degrees hotter than the darker lanes between them. The solar surface is in constant seething motion, rather like a boiling liquid, and each granule is an eruption of hot gas which then falls back again as it cools.

fundamental period of 2 h 40 m, indicate that they penetrate deep below the photosphere, and involve a substantial part of the interior mass of the Sun. Their periods then depend on the distribution of density through the solar sphere. Furthermore, the standing wave patterns of the oscillation rotate with the Sun, so that a long series of observations may show us how the interior is coupled to the differentially rotating surface.

Finally, the evolution of the Sun itself provides a fascinating problem. During the normal evolution of the Sun its luminosity should, according to standard theory, have increased by about 40% since the Earth was formed. The problem is to explain why the Earth is not icebound, which would be the case if the Sun were only 10% less luminous than at present. An icebound Earth is resistant to melting, since ice reflects the heat of the Sun. It requires a far greater output of energy from the Sun to reach the present watery state, although once reached it is stable at the present level of luminosity. It seems that we do not yet understand the Sun well enough to answer this vital question.

13

The birth, life and death of a star

Ancient tradition placed the stars at fixed points on the surface of a sphere – the eighth sphere, outside the moving spheres of the Sun, the Moon, and the planets. Our modern star-maps similarly depict a two-dimensional sky, like the geographers' description of the Earth's surface. The idea of a three-dimensional Universe, with the Milky Way interpreted as our Galaxy and the spiral nebulae as distant galaxies far beyond our local stars, is now easily accepted. But what do we know of the fourth dimension, time, which reveals the history and the fate of the stars? The chronography of the Earth is written in the geological record of the rocks: a similar record of the evolution of the Universe is written in the stars.

The Earth is about 4500 million years old, and the Sun is evidently older. When and how was it born, and what will be its fate?

The Sun is an average, normal star. If we want to study average, normal people we take a random selection of the population, giving a spread of ages, and we can at once describe the progression from the cradle through childhood and adulthood to the grave. So we can with stars, with one major difference: their lifetimes differ very widely from each other, and we have to classify them not only by age but by the speed of their development. The classification proves to be simple: the more massive a star, the faster it develops, and the brighter it shines.

Our samples of population can be taken in various ways. We look first at associations of stars, in which similar stars were born at about the same time and have not yet strayed far apart. Most of the stars of the Plough are associated together, as shown by their common proper motion. The Pleiades are a closely associated group of similar young stars, as are the stars of Crux. These groups are the local kindergarten of our stellar population.

Now we turn to a much older group – a globular cluster. All the stars in this cluster are at approximately the same distance, so that their range of apparent brightness exactly represents the range of their actual luminosities. Now we see more than a range of 100 in luminosity among stars which must have been born at the same time. This range is due to the spread in their rate of development, which is due to their spread in mass. To classify stars according to mass, we need observations of their colour and luminosity, and a model of their structure.

Classifying stars

Simple observation tells us that colours of stars cover a wide range. In Orion, Betelgeuse is red and Rigel is blue, as is Hadar (beta Centauri) in the southern sky. The Sun is yellowish white. Colour, and the detailed spectrum that goes with it, depends on surface temperature. The hottest stars, class O and B, have temperatures of order 30 000 K; classes A and F are cooler; class G (including the Sun) has a temperature of around 6000 K and the red stars class K and M complete the sequence down to 2500 K.

Temperature is not, however, enough for our classification. A white-hot star can be normal, or it can be a giant, or a white dwarf. Their luminosities are very different. In a cluster they are easily distinguished, but otherwise we need to measure distance so as to connect apparent brightness with actual luminosity.

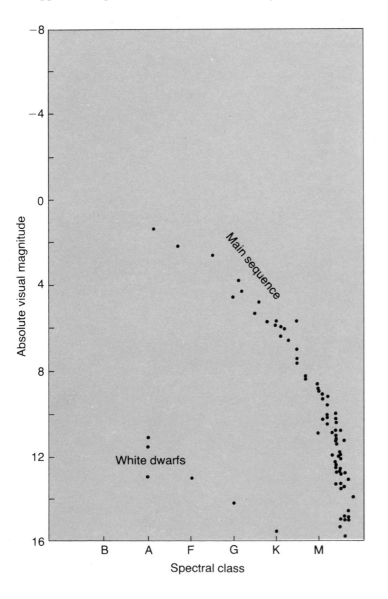

In a plot of absolute magnitude against spectral class in which each point represents an individual star, most of the points fall on a broad band, called the main sequence. This diagram has been plotted for a sample of the nearest stars. A plot of this type is called a Hertzsprung-Russell diagram after the astronomers who first employed the technique.

Let us now take a sample of a hundred stars close to the Sun, where we can easily see every star down to a very low luminosity, and we can find the luminosity of each from brightness and distance. Luminosity, expressed as an absolute visual magnitude, is plotted against colour, set out as spectral type: this is a Hertzsprung–Russell diagram. Apart from the white dwarfs in the lower left, the stars form a 'main sequence', with the hottest and most massive at the upper left and the coolest and least massive at the lower right.

A different sample, of the 100 brightest stars selected without regard to distance, is strongly biased towards the most luminous stars; only a few of the nearest stars (such as Sirius) appear in both samples. Instead of the faint K and M stars, and the white dwarfs, we now have superluminous stars known as giants and supergiants.

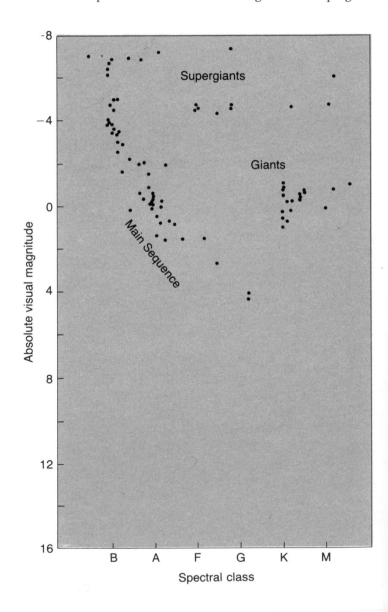

A Hertzsprung-Russell diagram for the brightest stars in the sky is biased towards intrinsically luminous stars. Only the upper part of the main sequence is represented and there is a high proportion of giant and supergiant stars, which are not common among the general population of stars.

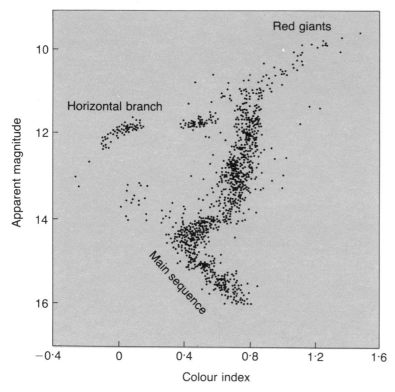

The distribution of points in a Hertzsprung-Russell diagram is characteristic of the particular sample of stars. In a globular cluster, which consists of old stars, the massive stars that used to populate the top end of the upper main sequence have evolved into giants. On this diagram colour index, a measure of the temperature or spectral type of a star, is used for the horizontal axis.

This is a passing short-lived phase, and only a small fraction of stars can be giants at any one time.

The connection between normal stars, giants and dwarf stars is found in our third sample, the old stars of a globular cluster. The spread of mass among these stars, all with the same age, shows as a spread along the main sequence. At the luminous, massive top end of the sequence, however, the stars have left the sequence and are moving up towards the red giant region. These are the rapidly evolving stars, and we will see that they leave the main sequence when they have begun to exhaust their energy supply.

Finally, the globular cluster again contains white dwarf stars, which are at the end of the evolutionary sequence. These are tiny stars, only about one-hundredth of the diameter of the Sun. The most famous is Sirius B, the almost invisible close companion of Sirius A. It is completely invisible to the naked eye, and it is very difficult to photograph as its light (visual magnitude 8.7) is swamped by that of Sirius A (visual magnitude −1.5). It became known first by its gravitational effect on Sirius A, since despite its small size it has the mass of a normal star.

These very varied stars fit together remarkably well in a simple evolutionary theory, which we now follow, starting with the condensation of a star out of the tenuous gas which is spread throughout our Galaxy.

The birth of a star

With the naked eye we see the Milky Way as a spectacular but patchy band of light stretching across the night sky in summertime.

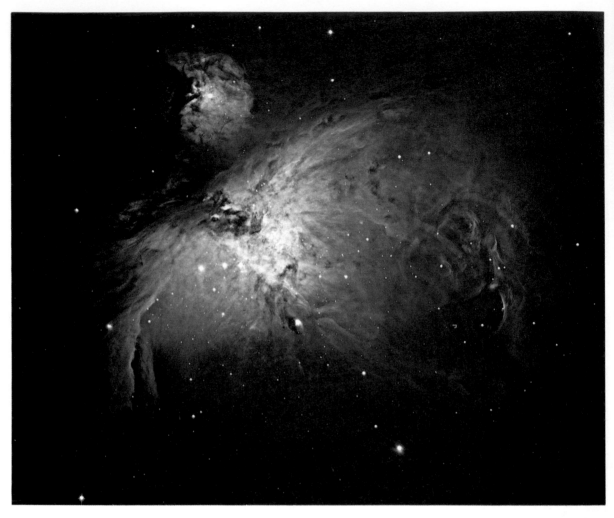

The Orion Nebula is a huge complex of dust and glowing gas in which is embedded a small cluster of very young, hot stars, known as the Trapezium. A special technique has been employed to create this photograph in a way that retains fine structural details.

It would be far more spectacular if the Galaxy were not filled with dust clouds.

Out of the dust clouds new stars are born by condensation under the simple force of gravity. The process is usually hidden from us by the dust cloud itself, but infrared astronomy has exposed some of the secrets of the star nursery. The best studied location is in Orion, close to the famous nebula.

The Orion Nebula itself is a hot cloud of hydrogen gas, glowing in the ultraviolet light of hot newly born stars at its centre. Behind the nebula is a cloud of dust and molecular gas known as the OMC (Orion Molecular Cloud). Infrared and radio maps of the region show up the OMC rather than the nebula. For example, the infrared satellite IRAS produced a map which may be compared with the optical view of the nebula. This infrared radiation comes from cold dusty gas, at a temperature of about 10 K.

The gas in the OMC contains an astonishing variety of molecules, which are detected by their characteristic spectral line radiation at radio wavelengths. These molecules form by condensation on the dust particles. More than 50 species of molecule have

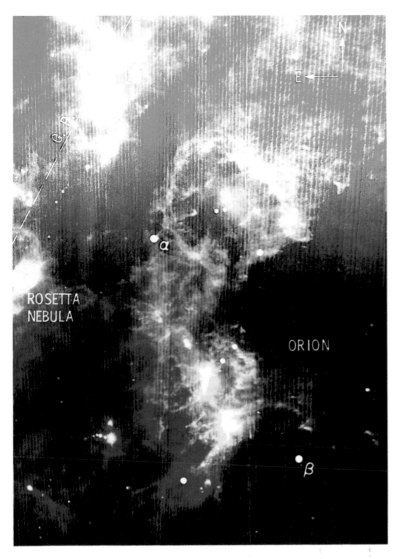

A region of the sky covering the constellation Orion mapped at infrared wavelengths from the IRAS satellite. Star images have been superimposed so the area can be identified and the Galactic Plane (the Milky Way) labelled G.P. The false-colour image of this region where stars are being formed shows strong radiation at 100 micrometres in red, at 60 micrometres in green and at 12 micrometres in blue.

been detected in the OMC and similar clouds. Some are simple: hydrogen, silicon oxide and water, while the more complex molecules include organic compounds such as alcohols and aldehydes. About one-half of the interstellar gas in the Galaxy is in the form of molecules.

Although the OMC is the birthplace of stars the molecules it now contains cannot be the basis of life. As soon as the condensation becomes a star, it becomes very hot and all molecules are disrupted.

Another visible birthplace is in the Carina Nebula. Some of the most massive stars have been observed here, notably η Carinae, which has about 100 solar masses. This star is unstable. In 1820 it brightened from magnitude 4 to 1, and in 1843 it reached magnitude -1, brighter than all other stars apart from Sirius. It then faded below magnitude 6, and is now slowly brightening again. Probably η Carinae has already run through its main sequence phase, and is heading for a supernova explosion a few thousand years hence. The whole history of such a massive star occupies only a million years; meanwhile other stars are forming right beside it.

Equilibrium – the normal stars

As a star condenses, its interior heats due to the release of gravitational energy. At a critical point, about 7×10^6 K, the temperature at the centre of the star is high enough for nuclear reactions to start. Hydrogen atoms start to fuse together in a sequence of reactions that create helium. An enormous supply of energy then becomes available, which is responsible for the light from the Sun and the stars: one gram of hydrogen generates 6.3×10^{11} joules of heat; hence, the steady conversion of hydrogen in the Sun can continue at the present rate for several billion years.

Why does the star not explode, but burn steadily for the whole of its lifetime on the main sequence? The equilibrium of a main sequence star is based on the simplest physical principle of equilibrium between the attractive force of gravity, which tends to collapse the star and raise its temperature, and the outward thermal pressure of hot gas, which tends to expand the star and cool it. The nuclear furnace is very sensitive to temperature: collapse under gravity increases temperature, increases the rate of nuclear reactions, which increases the thermal pressure, so the star expands and cools. The equilibrium state reached in this way depends almost entirely on the total mass.

Furthermore, the total energy production in the furnace is determined by the mass. Since all the energy must be radiated from the surface, the temperature of the surface must again be determined by the mass. Hence the neat organisation of the main sequence, as seen in the Hertzsprung–Russell diagram.

The equilibrium of a seesaw can be tested by giving it a small push and watching it rock about its balanced position. The equilibrium of a star can also be tested, since many of them show quiet but regular oscillations. These are the variable stars. The mechanism of oscillation is related to the diffusion of energy through the outer layers of the star; in some stars it causes a cyclic change of temperature, and in others a cyclic change of diameter. Both can be related to the structure and equilibrium of the star and its nuclear energy source.

We now see why stars have left the upper part of the main sequence in the Hertzsprung–Russell diagram. The cluster is so old that these heavy stars have used up the hydrogen at their centres. We now follow the more complex story of stellar old age.

The red giants

The helium which collects in the centre of the star after the hydrogen is exhausted is under tremendous pressure, but it is not yet hot enough to fuse into heavier elements. It therefore collapses, and the star shrinks. Hydrogen burning continues in a shell which works its way outwards through the star. The helium temperature eventually reaches 10^8 K, hot enough to start the next stage of fusion, when helium turns into carbon. Now we have a smaller star producing more energy; the consequence is that its outer layers of unburnt hydrogen are pushed outwards, forming a huge and cooler envelope which hides the hot core. The star has become a red giant.

Up to this point the mass had mainly determined the speed of evolution rather than its actual course. Now there is a divergence. The high mass stars follow the more spectacular paths which lead to a supernova explosion. Low mass stars, like the Sun, approach old age more quietly, with the core shrinking into a white dwarf made mainly of helium, carbon and oxygen. The outer envelope of the red giant is lost: it may be seen as an expanding nebula such as the Ring Nebula in Lyra. These nebulae are often called planetary nebulae, from a supposed resemblance to a planetary disc.

White dwarfs

No white dwarf star is visible to the unaided eye. Sirius B was photographed in 1917, before there was an explanation of a star which was so underluminous and yet so hot. New physics was needed to explain the hot, condensed material of a white dwarf, whose density is one hundred thousand times greater than the density of the Sun (which is not very different from that of ordinary solids and liquids). The physics is that of degenerate matter, like the physics of superconductors and superfluids which can be studied in low-temperature conditions on Earth. Complete gravitational collapse is now prevented by the pressure of a degenerate electron gas.

The high surface temperature of a white dwarf is a relic of its past, when it was still generating nuclear power. It is now destined to fade gently into oblivion, when it will become an unobservable black dwarf.

White dwarfs are the end-point of stellar evolution. Not surprisingly they account for about 10% of the total stellar population of our Galaxy.

Neutron stars and black holes

As we shall see in Chapter 16, stars with sufficient mass may not slide gently into old age, but may instead end with a supernova explosion. The core, if it survives the explosion, may now become a neutron star or a black hole. A core with more mass than 1.4 solar masses cannot be a white dwarf: the gravitational pressure cannot be resisted by the degenerate electron gas, and the star must collapse still further. The next stable state is a neutron star.

In the outer parts of a neutron star the pressure of gravity is resisted by tightly packed atomic nuclei. Inside, the nuclei merge into a sea of neutrons, and the density reaches one hundred million million times that of water.

The next stage is still speculation. There is an upper limit of about three solar masses for a stable neutron star. Above this mass further collapse is inevitable, and, if such condensations exist, they must form black holes. Here the force of gravity has finally triumphed completely, and the star has been crushed out of all observable existence.

The η Carinae nebula in the southern Milky Way is a complex agglomeration of gas and dust with bright hot stars at the core. η Carinae itself, at the centre, is a supermassive star that has declined dramatically in visual brightness since the middle of the last century but is the brightest infrared star in the sky.

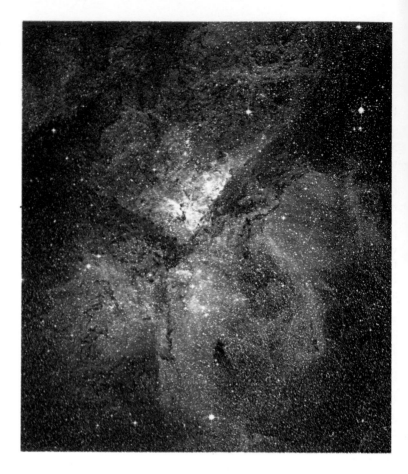

The time scale

The ticking of a Geiger counter, monitoring the decay of radioactive material such as radium, gives us a time scale for the youngest and most active materials on Earth. Older material, such as ancient granite or meteorites, can be dated by the slower decay of a potassium isotope. Chronography of the Earth is now well established, and we can date the formation of the Earth to about 4500 million years ago.

The Universe as a whole is about 14 billion years old, as shown by its rate of expansion. Where in this time scale can we place the birth and death of the stars in our Galaxy?

Most stars spend three-quarters of their lives on the main sequence: indeed, most of them have not left it. Their total available energy is known, and their rate of radiation is known. Their lifetime is therefore known. A very massive star, like η Carinae, can live for only a million years. The Sun can live for about 10 billion years. The least massive stars have scarcely started on the lower end of the main sequence; they cannot hope to exhaust their hydrogen inside another 20 billion years.

If we put these time scales alongside the population statistics, we find that about ten new stars are born every year in our Galaxy. The birth may be rapid: perhaps some tens of thousands of years for a dense dark cloud to light up as a new nuclear furnace. Death may be slow, as in the gentle fading of a white dwarf; or it may be spectacular, when a supernova explosion tears a star apart inside half an hour.

Binaries – a warning

The tidy theory of stellar evolution outlined in this chapter only applies to solitary stars. We shall see in the next chapter, however, that most stars are in binary systems. Let us not imagine, therefore, that we fully understand the story of stellar evolution in our Galaxy. How do binary stars condense? Do they usually have different masses? Do they usually interact with one another, so that if one becomes a red giant it can transfer mass to its smaller companion? The more detail we observe of strange stellar spectra, the more we see of neutron stars in strange orbits, and the more we see of gas flowing out of stars instead of falling on to them, the more we are forced to realise that we only understand part of a very complex story.

The Ring Nebula in Lyra is one of the brightest and most familiar of the planetary nebulae. The term 'planetary' refers only to the superficial resemblance to the disc of a planet since the nebula is actually a shell of gas ejected by a star in a late phase of its evolution.

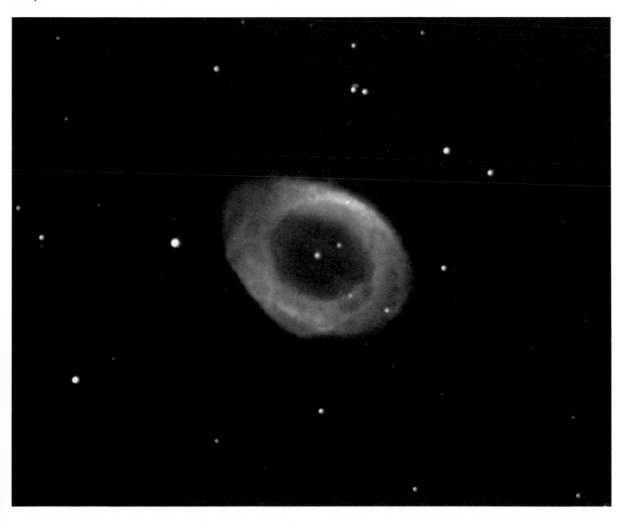

14

Binaries and clusters

The best way to start the study of astronomy is to choose a warm summer night, with plenty of time to spare, lie on one's back and gaze at the sky with unaided eyes. In the northern hemisphere the opportunity usually occurs during holiday time, say in August when there is the special attraction of the Perseid meteor shower. Away from the time of full moon, the great arc of the Milky Way, stretching from Sagittarius in the south up to Cygnus and Cassiopeia and over to Perseus in the north, can best be seen at this time of year. Overhead is the constellation of Lyra, just north of the Milky Way. The brightest star α Lyrae, is Vega. The fifth brightest ϵ Lyrae, is double, with two equally bright components 3.5 arc min apart. Check this with binoculars, and then try with the naked eye. It needs a clear sky for this to be possible, but it is just on the limit of angular resolution and you may not be able to separate the pair.

The binary ϵ Lyrae is a test for good eyesight. Closer binaries provide tests for good telescopes. At a good observing site the resolving power of a small telescope is, in principle, limited only by the diameter of its objective lens or mirror, so that a 20 cm diameter telescope should resolve a binary pair only 1 arc sec apart. But to reach a resolution of 1 arc sec a large magnification is needed, the optics must be very high quality, and a very steady mounting is essential. More modest targets for testing your skill and your telescope, or binoculars, are:

Southern hemisphere

α Centauri: separation 2–22 arc sec, period 80 years.
β Tucanae: separation 27 arc sec.

Northern hemisphere

α Geminorum (Castor): triple, separations 73 and 12 arc sec.
ζ Ursae Majoris (Mizar): separation 14.5 arc sec. Mizar also forms a wide pair (11 arc min separation) with Alcor.

Pairs like these may arise by the chance superposition on the sky of two stars at very different distances. The reality of genuine associations within these pairs is brought home by a startling fact: every one of the individual components of the binaries mentioned so far in this chapter is itself double. The classic is ϵ Lyrae, whose separate components, 207 arc sec apart, are doubles with separations of 2.3 and 2.8 arc sec. The two bright stars of α Centauri are obviously associated: they are in orbit round each other with a period of 79.9 years. There is a third component associated with α Centauri more than 2° away in the sky, slowly orbiting round the pair with a period of several million years: this is Proxima Centauri, a faint red star (magnitude 14.9) only 4.2 light years from Earth. Castor, or α Geminorum, the brighter of the Gemini twins, has three components each of which is a close binary. The periods are 9.22, 2.93 and 0.818 days. The first two pairs orbit each other with a period of several hundred years, and the third orbits at a greater distance, like Proxima Centauri but with a period of several thousand years.

A simpler binary, and one of the most glorious objects in the sky, is Albireo (β Cygni). This has yellow and blue components, magnitudes 3.2 and 5.4, 35 arc sec apart and just separable with good binoculars, such as 10 × 40, especially if they are mounted on a tripod.

Some stars, which cannot be resolved directly, are only known to be binary because their velocities reveal the orbital motion, through the Doppler effect on their spectra. These are the spectroscopic binaries. Some orbiting pairs have the plane of their orbit close to our line of sight, and one star may pass in front of the other: these are the eclipsing binaries. Some binaries were unseen or unsuspected until they were detected as X-ray sources.

With all these possibilities, it may be asked if there are any stars at all which are not in binary or multiple systems. The best example is the Sun: if this were in a binary system the planets would be thrown out of their orbits and we would not be here. There is of course a problem of definition: in principle all stars affect all others gravitationally, but a binary is not worthy of the name unless it is close enough for an orbit to be maintained for several revolutions. About three-quarters of the stars are then found to be members of binary or multiple systems.

Defining an orbit

It is a slow and unrewarding task to follow a visual binary round its orbit, as for example the star Krueger 60 was followed by Barnard. To be a visual binary the pair of stars must be well separated, and the period must be correspondingly long; only about a dozen visual binaries have periods less than 25 years. In contrast Algol, or β Persei, is an eclipsing binary with period 2 days 20 h 49 m; similarly β Lyrae has a period of 21½ days. The light curves of these can be followed by comparing their brightness with

The double star Krüger 60, in the constellation Cepheus, has a period of 44½ years. These photographs were taken in 1908, 1915 and 1920 by E.E. Barnard at Yerkes Observatory. The orbital motion is obvious. The stars have apparent magnitudes of 9.8 and 11.4 and are separated by about 2½ arc seconds.

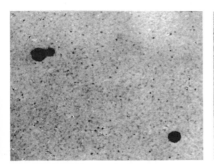 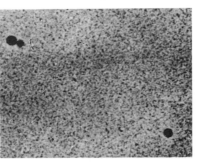 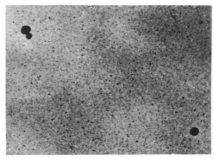

neighbouring stars. The patient observer may like to wait for an eclipse in ζ Aurigae, which occurs every 972 days.

The dynamics of a visual binary provide a direct measurement of the total mass of the two stars. Only the period and the distance are needed. The angular movements can be converted to a mean orbital radius A. Extending Kepler's law of periods for planets moving round the Sun, the binary mass is given by

$$\frac{m_{\mathrm{binary}}}{m_{\mathrm{sun}}} = \frac{A^3}{P^2}$$

where A is measured in astronomical units and P is the period measured in years. This the only direct way of finding the masses of stars, and it is a fundamental step in astrophysics.

The division of mass between the two component stars is not found so easily. If they have equal mass they will move at the same orbital speed, which may be measurable by the Doppler shifts of their spectral lines; if their masses are different, the ratio can be found from the ratio of the two Doppler shifts. Unfortunately this is not usually possible for the slow-moving members of the long-period visual binaries. Another parameter of the orbit that is hard to determine is the inclination of the orbit plane to our line of sight. A circular orbit may be seen as an ellipse, or even as a straight line: the binary γ Virginis, with a 180 year period, is steadily closing up, and will appear as a single star in 2016 AD. Even if the orbit plane is inclined to the line of sight it is often possible to find how nearly circular is the orbit itself from Doppler measurements round the orbit; in an elliptical orbit the velocities of the two components vary in a characteristic manner determined by the ellipticity.

In the classic case of Sirius, which is now known as a visual binary, only one of the pair was known when Bessel in 1844 showed that this bright star was pursuing a sinuous path across the sky, with a 2 arc sec deviation from the straight line of its normal proper motion. He deduced that it must have a fainter partner, Sirius B, which was photographed by Alvan Clark in 1862. It is almost ten magnitudes fainter than Sirius A, and it was only understood later to be a white dwarf, with a density more than 3000 times greater than that of its companion.

The masses of individual stars, which have been determined from their orbits in binary systems, range from 0.07 to 32 times the mass of the Sun. Many attempts have been made to look for planets, by searching for very small movements of their possible parent stars. A good candidate is Barnard's Star, in which there is a wobble of about 0.04 arc sec, currently interpreted as the effect of two planets with periods of 12 and 16 years, both with mass similar to that of Jupiter.

Close binaries

The eclipsing binary β Lyrae consists of two large stars, of 19 and 15 solar diameters, which are so close together that they are actually in contact. This accounts for the slow smooth variation in the eclipse curve, contrasting with the sharper events in the light-curve of Algol (β Persei); both of these are easily observed by the simplest of methods. The contact binaries are by far the more interesting class, as we can see in them stellar evolution in its most obvious form.

All the members of any individual binary or multiple star groups were formed at about the same time, so that they have the same age but they may have different masses. More massive stars burn up

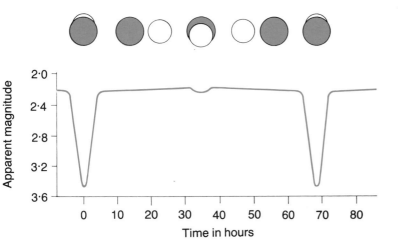

The light variations of Algol. There is a sharp dip in the total brightness of the system when the dimmer star eclipses the brighter one.

In the β Lyrae binary system the stars are distorted by their mutual gravitational attraction and the variations in the light curve are more gradual than in the case of Algol.

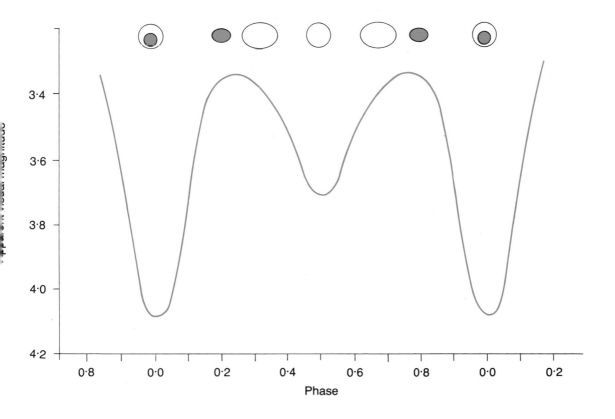

When one star in a binary system evolves and expands, it is drawn out towards its companion, filling what is called the 'Roche lobe'. The material falls onto the companion, forming an accretion disc around it.

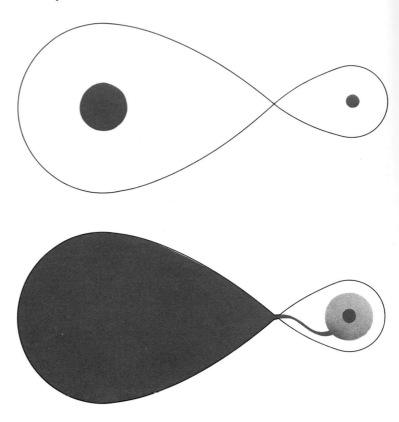

their nuclear fuel and evolve more rapidly. We will follow this process in detail in Chapter 16. When a star has burnt up its hydrogen it expands, and if it is sufficiently massive it becomes a red giant. It may have a less massive binary companion which is so close that it becomes enveloped in the outer part of the red giant, and this thin stellar atmosphere will fall on to the companion. The dynamics of this accretion process can be complex. The flow actually starts when the massive star's envelope is larger than a critical surface known as the Roche lobe. The accretion may also be in two stages: first the in-falling matter goes into orbit round the companion, and it falls in later, often guided by a magnetic field. Gravitational and orbital energy is dissipated during accretion. Algol emits detectable radio waves because of this extra energy; some particularly energetic accreting binaries become powerful X-ray sources.

Interaction of this kind occurs, or will occur, in about half of all binary star systems. It is, however, a short-lived phase in their evolution, and we only see it in a few special classes of stars such as the X-ray binaries.

X-ray binaries

The first X-ray stars were glimpsed for a few minutes only from sounding rockets and balloons, and their association with binary stars was not obvious until more continuous observations became possible with the UHURU satellite, launched in 1970. The X-rays from the source Centaurus X-3 (Cen X-3) were then found to disappear regularly every 2.087 days, which was a sure sign of a binary system. The eclipse was very abrupt, showing that a large star with a sharp edge was occulting a very small X-ray source. Then it was found that the X-rays were in pulses, with the very short period of 4.8 s. From previous work on the radio pulsars (Chapter 16), it was immediately obvious that the very small object was a rapidly rotating neutron star.

Here was a wonderful new opportunity to apply the classical binary star analysis. The period of 4.8 s varied through the 2.087 days of the orbit, which is a Doppler effect similar to that in the spectroscopic binaries, giving a description of the orbit complete except for the usual difficulty of knowing the inclination. The eclipse gives the final clue: for this to happen the plane of the orbit must be within 10° of the line of sight. Furthermore, the companion star has been identified optically; from its type and luminosity we know that it has a mass 20 times that of the Sun, and it follows that the mass of the neutron star emitting the X-rays must be between 1½ and 2 solar masses.

About a dozen X-ray sources like Cen X-3 have been investigated in similar detail. They all seem to be short-lived objects, as the accretion is so fast that it could not last more than 100 000 years. Their origin, and their probable fate, involves the story of supernovae explosions, which we deal with in Chapter 16. Our account of binary stars is, however, incomplete without a mention of another X-ray binary, Cygnus X-1.

Cygnus X-1 is visible optically, and as a radio source. It is very variable, as though it were being powered by a series of explosions. It is a binary, and the present interpretation is that the condensed object emitting X-rays is more massive than is possible for a neutron star. For this reason it is reputed to be a black hole; it may be the first real example of this remarkable concept.

Binaries can evidently be made up of almost any pair of partners that can be imagined, including an example of two neutron stars in orbit round one another, as described in Chapter 16 on pulsars.

Star clusters

In the northern hemisphere, everyone who knows any of the stars will recognise the Pleiades, a group of stars in Taurus recorded in classical literature. Most people can distinguish six stars, although the ancient references are to seven sisters or seven doves, and nine are named. It was one of Galileo's great discoveries, with the newly invented telescope, that there are many more: he counted 36, and there are about ten times as many to be seen with modern telescopes. The cluster is a fine sight in binoculars.

The keen-sighted observer, using a small telescope, may see that the stars of this cluster are surrounded by a blue nebulosity. This is starlight, reflected in a cloud of cold dust. The dust gives a clue to the origin of the cluster: it is the remains of a nebula out of which the stars condensed about 50 million years ago. All the stars are of the same age, but they show a wide range of brightness and colour, from bright blue to faint red, corresponding to the range from most to least massive.

The Pleiades form a young open star cluster surrounded by nebulosity that shines by reflected starlight. The cluster, in the constellation Taurus, is easily visible to the naked eye.

Another cluster easily seen by the naked eye is the Hyades. It is much nearer than the Pleiades, so it covers a larger area of sky and is less obvious. Look for the concentration of a few dozen stars covering about 8° of sky close to Aldebaran. The Hyades cluster provides an important stage in the scale of distances within the Galaxy. All the stars of the cluster are moving together at 45 km s^{-1} relative to the Sun. They are so close to the Sun that their tracks are spread widely across the sky, and the varying perspective gives an apparent variation of motion across the cluster. This variation depends on the distance, so that a measurement of apparent velocity for a set of cluster members gives the distance of the cluster from the Sun. The distances of other clusters can then be found from the ratio of apparent brightness of individual stars and their counterparts in the Hyades cluster: it is of course essential that the same types of stars are compared in this way so that they have the same intrinsic luminosity. Fortunately the Hyades cluster contains a good range of stellar types, including some old stars which have passed the stage of hydrogen burning.

Among the many other star clusters which are worth seeking out in a small telescope are the Double Cluster in Perseus (the two components are 1° apart, centred on the stars h and χ Persei), and the Jewel Box in Crux (centred on the star κ Crucis of the Southern Cross). These are all galactic clusters, which usually lie in the plane

Opposite The Jewel Box, a magnificent open star cluster in the southern constellation Crux, got its name from Sir John Herschel's description of it.

of the Milky Way. The globular clusters, in contrast, contain many more stars, very little gas and dust, and are located away from the plane of our Galaxy. The examples of galactic clusters which we have discussed are local objects, all much closer than the centre of the Galaxy: globular clusters can be seen at great distances, distributed through a spherical halo surrounding the Galaxy.

Globular clusters

Would-be observers of globular clusters should live in the southern hemisphere. Not only are the two most spectacular specimens, ω Centauri and 47 Tucanae, invisible from the northern hemisphere, but the vast majority of the 100 or more listed globular clusters can only be seen from the south. They are, in fact, the best indicators of the size of our Galaxy and of our place in it. A plot of their positions on the sky shows that they are widely distributed and not concentrated in the Milky Way, but centred on Sagittarius, where we now locate the centre of our Galaxy. The globular clusters form a spherical halo, and we are located at its edge.

The fortunate southern observer will find the brightest globular cluster ω Centauri, about 10° north-east of the Southern Cross, and not far from the Milky Way. Photographs of this cluster include a considerable number of foreground Milky Way stars, although the members of the cluster are overwhelmingly numerous. The next most famous example, 47 Tucanae, is well away from the Milky Way, and photographs show more clearly how the cluster population fades slowly away with increasing distance from its centre.

Northern hemisphere observers must be content with the globular cluster M13 in Hercules, just visible to the naked eye; this was listed as a nebula by Edmond Halley, and by Messier (whose initial

Although its name suggests that it is a single star, ω Centauri is in reality the largest globular star cluster in the Milky Way. To the naked eye it appears as a hazy star but actually contains hundreds of thousands of individual stars.

M appears in his catalogue of nebulae). M13 and its neighbour M92 are worth observing in binoculars, but a 20 cm telescope at least is needed for resolving separate stars near the centres.

The stars in these clusters cannot be counted individually, but one can work from the total brightness on the assumption that the star types are the same near the centre as on the outside. A typical population is between 10^5 and 10^6 stars, packed more than ten times more tightly than the stars in galactic clusters like the Pleiades. Even so, the stars are far from touching one another: if we represent a star by a tennis ball its nearest partner would be about 160 km away.

Globular clusters are remarkably well isolated from each other and from the rest of the Galaxy. Their average distance apart is about 10 000 light years, and they spend most of their time well away from the plane of the Galaxy. They are, however, bound to the Galaxy by gravity, and they follow orbits which take them right through the plane. ω Centauri has orbited the Galaxy more than 100 times during its lifetime, following a path which combines an oscillation through the plane with a retrograde motion round the centre, i.e. in the opposite direction from the general rotation of the Galaxy.

Globular clusters are old, probably as old as any other part of the Galaxy. They are fossil relics of the initial condensation of stars out of an original cloud of gas and dust. The outer parts of this cloud condensed into separate cloudlets which remained separate from the main spiral cloud of the Milky Way. Stars then condensed within the cloudlets, all within the same era, and we now see those stars as a remarkably uniform set, all belonging to the same generation. Globular clusters do not contain gas and dust, out of which new stars can be born: it seems that they are swept clean each time they pass through the Milky Way, removing any chance of a new generation of stars. We see a population of ancient star types, mostly faint red stars, including some white dwarfs. There has been very little of the process of synthesising heavy elements which we now associate with the activity of the Milky Way.

The uniformity of the star types allows us to use the globular clusters to estimate distances. In 1919, when the nature and distances of spiral galaxies like Andromeda were still unknown, Harlow Shapley showed that the spherical halo of globular clusters, on whose edge we are located, was centred at a distance of about 30 000 light years (his original value was about twice as much, but was corrected later (in 1930) by R.J. Trumpler who found that some correction was needed for absorption of the light from the more distant clusters). The best current estimate of the distance of the Andromeda Nebula was about half of this new measurement of the distance to the galactic centre. The revision of the distance to the Nebula came soon afterwards, as we shall see in Chapter 18.

The dynamical behaviour of the interior of a globular cluster can only be understood by measuring the velocities of a large number of its constituent stars. It is obvious, however, that the dynamics of a cluster are quite different from those of the Milky Way. There is some rotation, giving a small ellipticity, but there is no suggestion of a disc or a spiral. This difference is due to their great age and to the high density of stars. The mutual gravitational forces acting

A search for pulsars in globular clusters has been rewarded by success. A pulsar with a period of 3 milliseconds and in a binary orbit was discovered in 1987 in the globular cluster M4 at the location shown by the white cross.

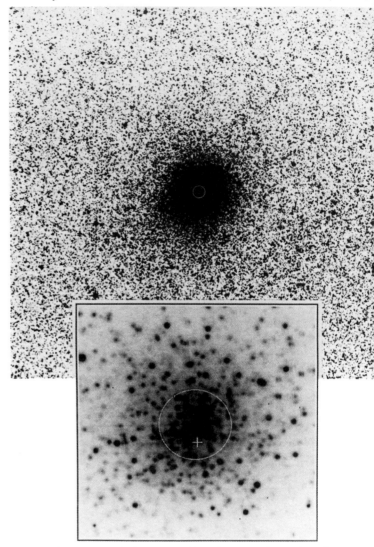

over a long period have produced a random distribution of velocities rather like the kinetic velocities of gas molecules. This settling down process, known as relaxation, involved the loss of a proportion of the highest velocity stars which carried away most of the original rotational angular momentum.

The centres of globular clusters are hidden from us by the tight concentration of stars. There is a possibility that many stars could have coalesced in the centre, forming a very heavy star or even a black hole. Another possibility is that close binaries are common in the centre. Interest in such speculation has been sharpened by the discovery of several X-ray sources located in globular clusters; one is in NGC 6624 and another in M15. Not much is known of these X-ray sources, and long observations are needed to search for binary periodicities or short pulsations which might give clues like those to the nature of the X-ray binary Cen X-3.

Two short-period pulsars have been discovered in the globular clusters M4 and M28; these are the older type of pulsars, known as 'millisecond' pulsars. The one in M4 is in a binary system.

15

Variables and novae

We usually attribute to Aristotle the view that the heavens are unchanging, apart from the organised motions of the Sun, Moon and planets. The stars used to be regarded as a permanent and perfected arrangement of bright points on a hemispherical surface. Chinese astronomers in the tenth century AD were much more open to the possibility of change; their records contain many interesting references to comets, meteors and new bright stars. There are very few such references in western records, until Tycho Brahe's observation in 1572 of a new bright star, a supernova, in the constellation of Cassiopeia. The realisation that the brightness of stars could vary periodically dates from 1596, when Fabricius noted that a star in the constellation of Cetus varies with a period of 11 months. He named it 'Mira', meaning 'the marvellous'. Mira reaches magnitude 2 at its brightest, and disappears completely from naked eye observations at its faintest.

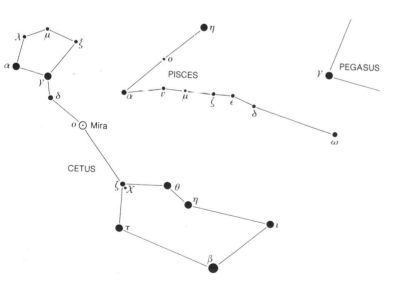

The location of the variable star Mira (o Ceti) in the constellation Cetus.

The light curve of Mira showing the variation over six magnitudes in a period of around eleven months.

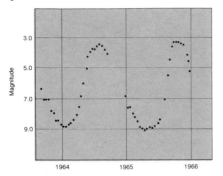

In contrast to the Aristotelian view, we now know that at least one-third of the bright stars are variable, although some vary only over a very small range. Leaving aside the eclipses in binary systems, the variations are of several types. The star R Corona Borealis falls from magnitude 6 to magnitude 13 for about a month at intervals of around 5 years: this is the effect of an obscuring cloud condensing round the star. U Geminorum occasionally brightens from magnitude 14.3 to magnitude 9, and fades again within a few weeks: this is a dwarf nova and is in a binary system. Many stars, like the Sun, vary slightly as they rotate, because of an irregular pattern of spots on the surface. But the most interesting of the variable stars are those which have a built-in and precise periodicity due to an internal oscillation. Of these the classic case is the star δ Cephei, which has given its name to the Cepheid variables.

Observing variable stars offers an interesting challenge to the amateur. For some of the less predictable and long-period variables amateurs can perform a service by monitoring their behaviour and warning the big-telescope observers when an interesting phase has been reached. There is also a dedicated band of amateur sky-watchers who are usually first to pick up novae and supernovae.

Mira itself is a difficult variable to follow. Its period is close to a year, so that its bright phase will occur persistently for some years when it is close to the Sun, while the faint phase is too faint for small telescopes. A guide to easier examples is given by most amateur astronomical societies, and particularly by the British Astronomical Association and by the American Association of Variable Star Observers.

The Cepheids

The prototype of this class of variables, δ Cephei, was noted by Goodricke in 1784. It varies from 4 to 5 magnitude with a period of 5.37 days. The Pole Star, Polaris, is also a Cepheid, with a period of 3.97 days, and a very small range of brightness, between magnitude 2.08 and 2.17.

The fame and the usefulness of the Cepheids stems from the work of Henrietta Leavitt, who in 1908 observed a number of them in the Small Magellanic Cloud. This is a nearby galaxy, sufficiently compact for it to be assumed that all the Cepheids were at the same distance. Their range of apparent brightness therefore corresponded exactly to their intrinsic luminosity, or absolute magnitude. Miss Leavitt found a remarkably close relationship between the period and luminosity of the Cepheids; this meant that the absolute luminosity of any Cepheid could be found merely by measuring its periodicity, and its distance then followed from a measurement of apparent brightness. We will see that this forms an important stepping-stone in the distance scale of the whole universe (Chapter 18).

This remarkable relationship can be explained very simply. The periodic change in brightness in a Cepheid variable is due to an oscillation in the star, when it swells and shrinks like a heartbeat. The oscillation period is the result of an internal resonance, determined by the time taken for a compression wave to travel from the centre to the surface. This depends on the mean density, just as a

sound wave depends on the density of the gas in which it travels. (This is why one's voice becomes high pitched and squeaky if one takes in a deep breath of helium gas.) The structure of all Cepheids is similar, so that larger, more luminous Cepheids are more dense, and resonate with longer periods.

One further piece of astrophysics is needed: how is the oscillation

The Small Magellanic Cloud is a compact galaxy neighbouring our own Milky Way. The study by Henrietta Leavitt of Cepheid variable stars within it resulted in an important technique for establishing the distance scale of the Universe.

Cepheid variables, which are unstable pulsating stars, show regular changes not only in brightness but in temperature, radius and the velocity measured along the line of sight.

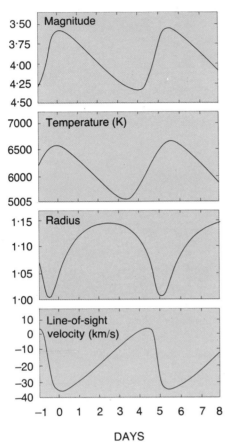

The light curve of a typical RR Lyrae variable. As with Cepheids, the variability of RR Lyrae stars arises from pulsations within them.

excited? There is plenty of energy available from nuclear fuel at the centre, but instability of the outward energy flow needs some sort of valve mechanism. This is provided by a layer in which hydrogen and helium are partly ionised. When this layer is compressed it becomes more opaque, trapping radiant energy inside. The pressure then expands the star, the valve opens, releasing radiation, and the star contracts again. There is very little effect at the centre of the star, where the energy is generated, but the surface pulsates in and out with velocities of some tens of kilometres per second.

The variation of brightness in a Cepheid is mainly due to temperature change at the surface, and only partly due to the change in diameter.

W Virginis variables

The class names of variables are usually derived from the prime example of each type, a practice which offers little help to the uninitiated. W Virginis is one of a list of variables in the constellation of Virgo, the letters running from R to Z. Past Z the letters then run from RR to RZ, SR to SZ etc., and then from AA onwards.

W Virginis variables behave very similarly to the Cepheids, but their luminosities are about four times fainter for a given period. They were once confused with Cepheids, with the result that some distances to globular clusters and extragalactic nebulae were overestimated by a factor of two. They are sometimes called Type II Cepheids, perhaps to commemorate this confusion, but they are, in fact, at a quite different stage of stellar evolution; they are comparatively old, low mass stars, which have just reached the stage of helium burning. They are in a state of rapid evolution, so are comparatively rare. One of them, RU Camelopardalis, actually stopped oscillating in the early 1960s, presumably having evolved from the rather critical conditions needed for instability.

RR Lyrae variables

These are common in globular clusters; they belong to the popula-

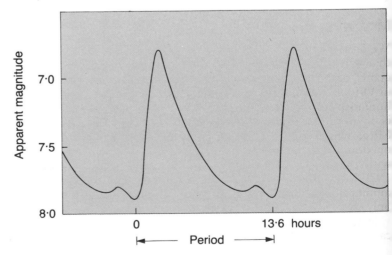

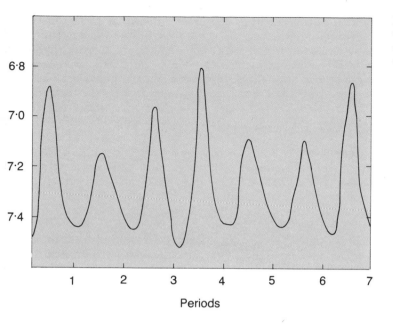

Periods

In SX Phoenicis, two modes of pulsation take place simultaneously so that the amplitude of the main variations changes regularly.

tion of old stars forming the halo of our Galaxy. They have short periods, from 1.5 h to 1 day, and a range of brightness of 0.5 to 1.5 magnitude. Like the W Virginis variables, they are stars at the beginning of the helium burning stage, but they are giant stars, larger and hotter and therefore more luminous. Unlike the Cepheids they all have the same intrinsic luminosity; this means that, like Cepheids, they can be used as 'standard candles' in estimating distances.

The light curves of RR Lyrae variables are of two types depending on the nature of the resonance in the star: fundamental or overtone, as in a violin string bowed in different ways. A related class of variables, the Dwarf Cepheids, also have two modes of resonance, and in some, such as SX Phoenicis, both modes are excited at the same time. The result is a beating effect.

Mira variables

We now see that the period of 11 months for Mira must indicate that it is a very large star. All the long period variables are red giants, and very bright. Betelgeuse itself, the outstanding red giant in Orion, is variable, but only irregularly variable. Variations in luminosity of a giant star might be due to a change in diameter. This is obviously insufficient by itself to account for such a variable as Mira, whose visible light varies by a factor of 1700 between maximum and minimum. In fact the variation is mainly associated with the surface temperature, which changes from 2640 K to 1920 K, so that the peak luminosity moves into the infrared at minimum. The total energy output (the 'bolometric luminosity') only changes by a factor of four.

Mira itself is a double star. Its companion is a white dwarf which can be seen clearly when Mira is at minimum, so that the orbits and masses of the binary pair can be determined. We know therefore that Mira is a star of one solar mass and has several hundred times the luminosity of the Sun. Oddly enough, the white dwarf itself seems to be variable: perhaps this is not so surprising when one contemplates the number of different ways in which instability can occur, and the large proportion of stars which are variable.

Are there any non-variable stars?

It seems that the more we know about stars, the more of them appear to be variable or unstable in some way. We have already mentioned the Sun as a rotational variable: it also vibrates in many different modes simultaneously, although this alone scarcely qualifies it as a variable. But many of the brightest stars in the sky, the apparently stable hot massive stars such as β Cephei, and β Canis Majoris, α Virginis (Spica), β Crucis and β Centauri, are now known to be slightly variable with periods of a few hours. The luminosity range is only a few hundredths of a magnitude, and the mechanism is unknown.

Novae

On 29 August 1975 a young astronomer, Paul Cass, was setting out for his night's duty as a site-tester on the proposed site of the Roque de los Muchachos Observatory on the island of La Palma, in the Canary Islands. He glanced as usual over the evening sky and immediately noticed an extra member in the constellation of Cygnus. He was among the first of many European discoverers of Nova Cygni 1975: others included Tucker, the duty observer at the Royal Greenwich Observatory, and Patrick Moore, who by chance was carrying out a routine series of observations of variable stars in Cygnus, and found his telescope pointed straight at a new bright star. All these discoverers were, however, preceded by a few hours by Japanese observers, who seldom miss such an opportunity.

The light curve of Nova Cygni 1975 shows the rapid rise and decline over a few days, followed by much slower fading.

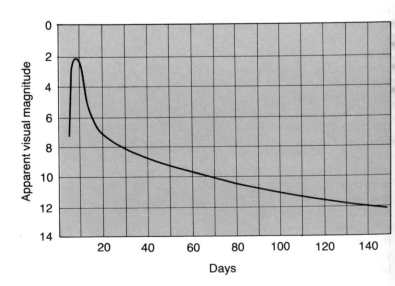

Nova Cygni 1975 rose to above magnitude 2 in a few hours. Sky survey photographs of its location before the outburst show nothing brighter than magnitude 21. This ratio of luminosity is at least 40 million! The light curve shows that it remained visible to the naked eye for only a few days. The spectrum showed that the light originated in a shell of material ejected from the star, travelling outward with a velocity of 1000 km s^{-1}. This shell became an emission nebula, with emission spectral lines typical of low density gas. A similar spectrum can be seen in a planetary nebula, which is a cloud of gas thrown off during the collapse of a star to become a white dwarf. In fact, the typical nova also involves a white dwarf, but as a member of a binary system rather than as a solitary evolving individual.

The binary nature of novae often shows up in the later stages of the outburst. Some are eclipsing variables, in which a remaining flickering light is periodically eclipsed by an unseen companion. These are very close binaries, in which a white dwarf is pulling material from a larger companion. The in-falling material collects in a ring round the white dwarf; the ring can be detected in X-rays and ultraviolet light, and it also emits a flickering light from the main seat of accretion into the ring.

The nova phenomenon is at the surface of the white dwarf rather than in the accretion ring. The white dwarf itself has passed the youthful stages of life when it can generate energy from nuclear burning. Hydrogen atoms from its companion fall on to its surface,

The 1975 nova in Cygnus at its second-magnitude maximum and after declining to fifteenth magnitude. The outburst occurred as the star threw off an outer shell of material.

providing fresh fuel. When sufficient fuel has accumulated, a massive nuclear explosion occurs, sending material flying outwards in an expanding shell. Accumulation can then start again, allowing a recurrence after some hundreds or thousands of years. In the so-called 'dwarf' novae, the explosion and the ejection are less drastic, and the cycle can recur in only a few years.

Recurrent novae are generally predictable but not precisely so. The recurrent nova T Coronae Borealis is expected again in 1993; it will be the object of nightly attention by observers anxious to see the first stages of this massive nuclear explosion.

X-ray novae

The satellite X-ray observatories show us a far more variable sky than we are used to seeing optically. Stars must be very hot indeed to emit X-rays; temperatures of a few million degrees rather than a few thousand must be involved. A white dwarf in a binary does not get hot enough, but the surface of a neutron star, small though it is, can be hot enough to emit detectable X-rays when it is accreting matter from a binary companion. This happens either in a steady process, as in the pulsating X-ray sources (which are actually rotating neutron stars with a hot spot on the surface), or catastrophically to produce an X-ray, or even a gamma-ray, transient.

As for the optical novae, the process starts with hydrogen accreting from a cool giant star on to the neutron star. The energy of the fall under gravitation is already very large: a peanut falling on to the surface would release energy equal to that of a small nuclear bomb. But a further effect is now believed to account for the X-ray novae which burst out occasionally. The accumulated hydrogen on the surface can itself explode like a nuclear fusion bomb, flashing up to temperatures of perhaps 10^8 K within a fraction of a second.

Flare stars

We turn finally to another group of irregularly variable stars, whose instabilities are intrinsic and not fed by a binary partner. These are the flare stars, often known as UV Ceti stars.

Flare stars brighten unpredictably by several magnitudes, then fade to their previous quiescent state in a few minutes. They are hard to observe, even though they are probably very common, as they are generally very faint. It is hard to devote sufficient patience and telescope time to catch and record these flares, which occur at intervals of some hours or days. They are, however, particularly interesting on account of the radio emission which can often be detected at the same time as the optical flare. Radio waves are emitted by the stellar atmosphere, and it seems that the surface disturbance which produces the light is propagated outwards as a shock wave to excite the radio emission. The Sun is a flare star, in the sense that it occasionally emits powerful bursts of radio waves. These originate in disturbances of the magnetic field near sunspots; there is, however, compared with the red dwarf flare stars, little change of optical luminosity and total energy output, which is fortunate for the inhabitants of Earth.

16

Supernovae and pulsars

One of the easiest northern constellations to recognise is Cassiopeia, where five bright stars form a conspicuous W. Anyone familiar with the sky would be astonished to see an extra star, brighter than the planet Venus, forming a dot over the centre of the W. When this actually happened, in 1572, the conventional view of the heavens was that the stars were fixed and unchanging: the new 'star' must be something local, perhaps in the Earth's atmosphere, or possibly in a planetary orbit, like Venus. The Danish astronomer, Tycho Brahe, therefore measured its position day by day in relation to the normal stars in Cassiopeia, and found that it did not move throughout the period of more than a year during which it was visible. It was indeed a star, and the ancient picture of the unchanging heavens was shattered.

Tycho Brahe was a model observer. Although the new star was lost to view after 15 months, the position he measured now marks the centre of an expanding bubble of gas, the remnant of the supernova explosion which he so carefully recorded. The bubble is faint, and can only be seen in long exposure photographs, but it does emit powerful radio waves and X-rays. Another supernova, following in 1604, was similarly pinpointed by Kepler; its faint remnant was found in 1943, and it is also a powerful radio source. Yet another, again in Cassiopeia, must have occurred in about 1670; there are no records of a new star at that time, even though several European observatories were starting regular observations, and we know of it only from the remnant. This is known as Cassiopeia A, the brightest radio source in the sky; it can be dated simply from its rate of expansion, which can be seen on photographs taken only a few months apart.

There are only a few other recorded sightings of supernovae in our Galaxy. The most famous goes back to 1054, when Chinese, Korean and Japanese astronomers recorded the birth of the Crab Nebula. The Chinese had no inhibitions about an unchanging sky; they were determined to record everything that happened: comets, meteors, new stars, anything that might be used to predict future events. They were astrologers, not astronomers, but their description of this supernova and its position in the sky forms the greatest scientific treasure in Chinese history. They were very thorough observers, working from specially constructed towers, where, according to an account by visiting Jesuits in 1696,

'Five mathematicians spend every night on the Tower watching whatever passes overhead; one is gazing towards the Zenith, another to the East, a third to the West, the fourth turns his eyes Southwards, and a fifth Northwards, that nothing of what happens in the four corners of the World may scape their diligent observation.'

This system had apparently been used without a break for the previous 16 centuries!

No supernova has been seen in our Galaxy since the invention of the telescope. Astronomers have, however, been able to use their modern telescopes and spectrographs on a supernova in a nearby galaxy, the Large Magellanic Cloud. This is at a distance only about five times that of the centre of our Galaxy, and the supernova of February 1987 (SN1987A) was easily visible to the naked eye. The star as it was before its explosion could be seen on earlier photographs, so that the whole process could be followed. The most exciting observation was, however, not optical; in fact it was made in the northern hemisphere where the Magellanic Clouds are always below the horizon. We know that much of the tremendous energy of the explosion must be released in the form of neutrinos. Detecting these is very difficult; they even pass through the Earth almost without interaction. It happened that three large detectors were in operation at the time of the explosion, all in the north. A total of nineteen neutrinos were detected, all of which must have already passed right through the Earth. The fact that they all arrived within a few seconds, after a journey of two hundred thousand years was almost as exciting as the detection itself.

No one can predict the date of the next supernova in our Galaxy, but it is likely to be an amateur with no more than binoculars who will be the first to see it. The chances of a supernova occurring are, however, very slim, as none have been seen since 1604. There have, of course, been novae, which brighten and fade within days. Novae

Two false-colour images of the shell of material expanding outwards from the supernova observed by Tycho Brahe in 1572. On the left is a computer-enhanced optical image from the US National Optical Astronomy Observatories and on the right a colour-coded map of radio emission at 1370 MHz from the Very Large Array in New Mexico

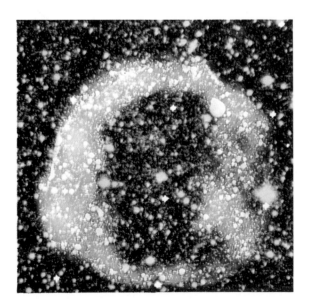
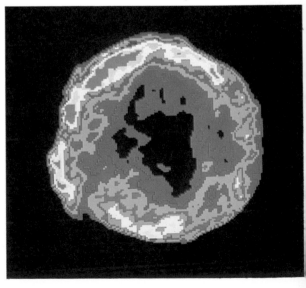

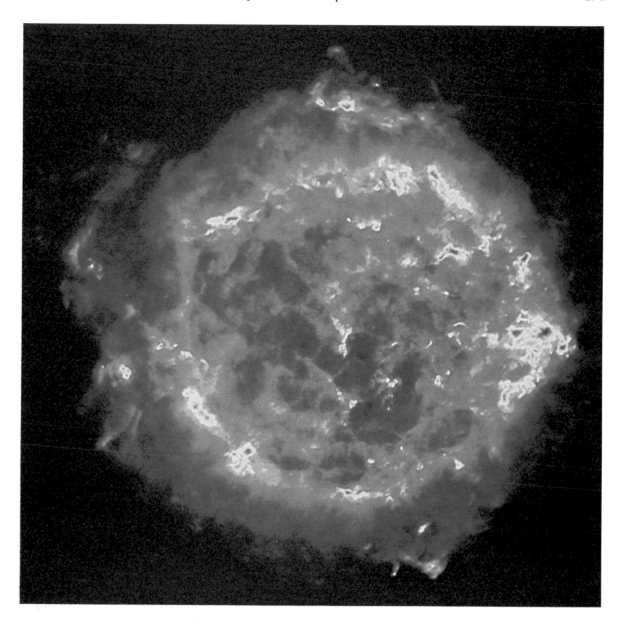

are exciting occurrences, when a star suddenly becomes more than one million times more luminous than the Sun, but the supernovae are ten thousand times more luminous again. They can be as luminous as a whole galaxy! It is fortunate that they are so rare that there is little chance of one occurring near to the Sun; if this happened, life on Earth could be completely destroyed.

At a safer distance, supernovae can be observed in other galaxies; telescopes can be used to photograph the same galaxy night after night. Some galaxies produce several supernovae every year. One was discovered solely by its radio emission.

Cassiopeia A is the remnant of a supernova that exploded about 300 years ago. This radio map was produced with the Very Large Array.

The discovery of a supernova in the Large Magellanic Cloud in February 1987 was one of the most exciting events in astronomy for many years. The lower photograph is the discovery picture obtained on the 24th February 1987 by Ian Shelton of the University of Toronto, who was working at the observatory on Las Campanas Mountain in Chile. The photograph above was taken just two days earlier. The arrow points to a blue supergiant star which is believed to be the one that exploded.

Collapse and explosion

The stupendous power of a supernova explosion tells us immediately that this cannot be a minor event on the surface of a star. The whole star must explode, producing within a few seconds more energy than our Sun has radiated in its whole lifetime. It is in fact the closing episode of the long evolution of a typical fairly heavy star, as we saw in Chapter 13. It is the last dying gesture of a star that has used all its fuel and has no further means of signalling its presence to us. How then can it summon up the energy to produce the most spectacular event in its whole history?

The driving force of a supernova explosion is gravity. Nuclear energy, from the fusion of hydrogen into helium, and helium into the heavier elements, keeps normal stars hot, and thermal pressure from this power source balances gravity. When the centre of a star has used up all its sources of nuclear power, it has a core of iron, surrounded by shells of lighter elements which can still produce some energy. The fusion of these elements keeps the temperature very high, and the whole star progresses steadily towards an inert mass of the final fusion product, iron.

Catastrophe comes swiftly, when the iron core itself disintegrates under the influence of gamma-rays from the shells around it. The iron breaks up again into helium atoms, which themselves break up into electrons and protons, and finally into neutrons. This is the reverse of the usual nuclear power source; the core now has no strength and no thermal energy to resist gravitational collapse. Unseen from outside the star, the core shrinks to an insignificant speck, and the unsupported shell around it immediately collapses. It is this shell that rapidly becomes visible, since the energy of the collapse heats it to temperatures of many millions of kelvin. Furthermore, the disintegration of the iron core releases high energy neutrinos, which add to the energy of the shell. The result is a rebound and the expanding bubble of gas which we now see. The outward speed is around $10\,000$ km s^{-1}.

This rare event is one of the most important in the Galaxy. When the shell collapses, it already contains some heavier elements such as oxygen, carbon and silicon, but none heavier than iron. In the explosion all the heavy elements, right up to uranium, are synthesised. Without these life could not exist. They are dispersed throughout the Galaxy, and eventually form part of newly born stars. One such star is the Sun; its planet Earth is our home thanks to prehistoric supernova explosions which provided the vital elements of our existence.

The Crab Nebula

The remnant of the supernova of 1054 is worth looking for, although it appears in a small telescope as no more than a fuzzy patch of light. It is visible from anywhere north of latitude 40°S. Look on a dark night, near new moon, and away from artificial lights, during a January evening (or near midnight in December, or evening in February). The nebula is approximately 1° north-west of ϵ Tauri. It is often referred to as M1, the first entry in Messier's catalogue of nebulae.

The Crab Nebula is still shining with the light of 30 000 of our Suns; it is also an emitter of radio, infrared, ultraviolet and X-rays. The source of energy for this continued activity, more than 900 years after the explosion of the central star, was until recently a complete mystery. Other remnants, such as Tycho Brahe's or Kepler's, are evidently decaying fast, using up the remains of the energy released in the supernova explosion. But the Crab Nebula has its own powerhouse still operating to generate high energy electrons and a strong magnetic field, which together produce most of the radiation we now see. That powerhouse is the Crab Pulsar.

The Crab Nebula is the remains of a supernova seen by Chinese observers in 1054. This is a composite of images taken in red light produced by hydrogen and blue light from sulphur. It shows up the structure in the nebula and how the physical conditions and chemical composition vary.

The pulsars

In the centre of the Crab Nebula there is a star rotating 33 times per second, sending out a narrow beam of light which flashes past us every 30 milliseconds, like a maritime lighthouse but more than one hundred times faster. The source of energy for the Nebula resides in the rotation of this star.

The collapse of the central core in a supernova is halted only by the close packing of neutrons. These pack so close together compared with normal atomic spacings that one solar mass, which is about the mass of a supernova core, packs into a sphere only 30 km across. The density is fantastically high: about a billion (10^9) tonnes per cubic centimetre. This core is left as a neutron star at the centre of the expanding supernova remnant.

The concept of an inert star only 30 km across is unattractive to an observer, who would not expect to see anything of it. Even if the surface temperature were 10^6 K, nothing would be visible from so small an object, and only the most sensitive satellite-borne X-ray telescopes would have a chance of detecting the thermal radiation. It is therefore a wonderful gift for astronomers that several hundred of these neutron stars make themselves known as pulsars.

A pulsar is a neutron star which emits a narrow beam of radio waves and, rotating like a lighthouse, sweeps the beam past an observer once per rotation. The rotation speed for some pulsars

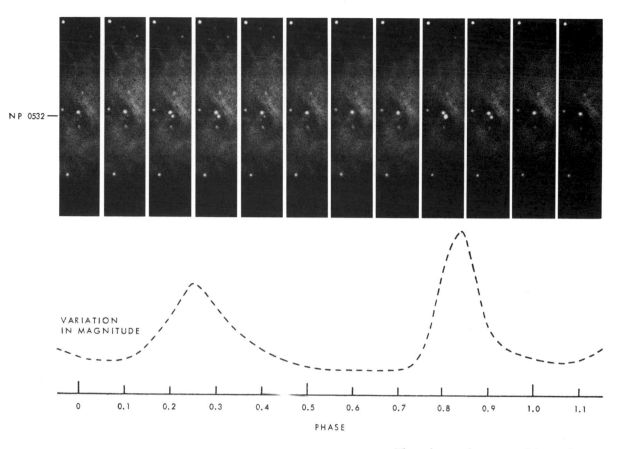

VARIATION IN MAGNITUDE

PHASE

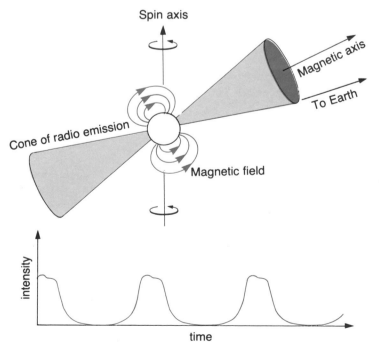

Spin axis

Magnetic axis

To Earth

Cone of radio emission

Magnetic field

intensity

time

The pulsar at the centre of the Crab Nebula flashes on and off thirty times a second. The cycle of light variation includes a sub-pulse and a main pulse.

The radio pulses from a pulsar are caused rather like the beam from a lighthouse. As the cone of radiation from the spinning neutron star sweeps past the observer's line of sight, the radio emission appears as a pulse, the shape of which depends on the orientation of the spin and magnetic axes of the pulsar.

approaches 1000 rev s^{-1}, giving a pulse period just over 1 ms; most pulsars have periods between 0.1 and 2 s. The Crab Pulsar and the Vela Pulsar in the southern hemisphere also radiate a light beam, and even a beam of gamma-rays. Neutron stars with similar behaviour, but rotating more slowly, are found as X-ray pulsating sources in some binary systems.

The pulsar powerhouse

The rapid rotation of pulsars is a direct result of the collapse of the supernova core. Just as the pirouetting skater spins faster by pulling in his arms, so does a slowly spinning star spin much faster by shrinking its radius over 1000-fold. Further, stars have a magnetic field, usually rather larger than the Earth's field: this also becomes enormously magnified when the star shrinks. The result is a gigantic dynamo, generating elementary particles with very high energy.

In the Crab Nebula, the energy generated by the pulsar dynamo is mainly transferred to the particles and magnetic field of the nebula. Only a small part feeds the lighthouse transmitter. But this transmitter is a wonder. Its output is detected over practically the whole of the available electromagnetic spectrum, from radio through infrared, visible and ultraviolet light, to X-rays and gamma-rays.

The pulsar clock

The rotation rate of most pulsars is very uniform, suffering only a very slow decay as the energy is radiated away from the pulsar dynamo. One of the 'millisecond' pulsars can be timed so accurately that its pulse arrival time is measurable to better than a microsecond, and at this level of accuracy no irregularity has been detected in some years of observation. Our standards of time on Earth are the atomic clocks, locked to the vibration of caesium atoms. It now seems possible that pulsars are better clocks than our present standards, although a proof would require a long series of observations of several of the most stable pulsars, using large radio telescopes. Meanwhile one of the pulsars, in a binary orbit round another neutron star, has already been found to be sufficiently stable to provide us with a new test of the predictions of relativity theory.

The relativistic binary

Most pulsars are solitary; they may once, like many stars, have had binary companions, but these have been lost at the time of the supernova explosion. The pulsar PSR 1913 + 16 is an exception: it is in a close, highly elliptical orbit round another neutron star. The orbital period is only 7 h, and the orbital speed is an appreciable fraction of the velocity of light. Here is an ideal testing-ground for relativistic theory: a clock moving very fast in a varying direction, and in a varying gravitational field. The first effect is the Doppler effect: as the clock comes towards us it seems to go fast, and it slows as it goes away. Next there is the transverse Doppler effect as it crosses the line of sight; these effects are both predicted by special relativity.

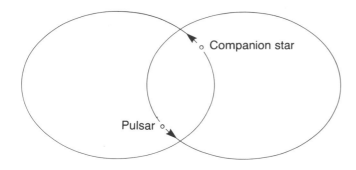

When a pulsar is a member of a binary system, the line-of-sight velocity and the number of pulses received from it each second vary systematically as the pulsar orbits around its companion.

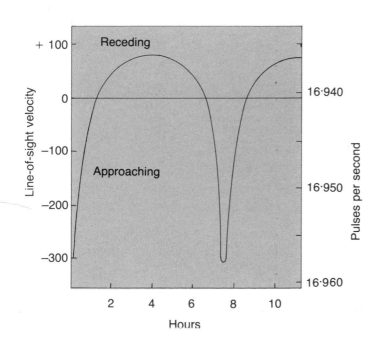

The most interesting effects are those predicted by *general* relativity, starting with the apparent slowing of the clock when it is most affected by the gravitational field of its companion. A major test of general relativity concerns the binary system as a whole. Two massive bodies revolving about each other should radiate gravitational waves. These are too weak to be detected on Earth, but the theory has nevertheless been tested by measuring the slow change in the binary orbit due to the loss of energy in gravitational radiation. Astronomy has again provided the only test for a fundamental theory of our physical world.

17

The Milky Way — our Galaxy

During the northern hemisphere summer (or southern hemisphere winter) the Milky Way paints a blazing trail over the night sky. It is easy to show, as did Galileo, that it is composed of myriads of stars. The simplest of telescopes or binoculars will suffice to resolve the brighter stars from the overcrowded background. But for the easiest demonstration that the Milky Way is a galaxy, our Galaxy, we must turn to radio astronomy.

The map of radio emission from the whole sky was put together from observations made in both hemispheres. Radio waves penetrate the dark interstellar clouds, giving us a view of previously unseen distant parts of the Galaxy. We are looking at a disc-like galaxy from a position some distance away from its centre. The galactic 'equator' marks out the plane of the disc and there are corresponding galactic 'poles'. But we still have no dimensions: what are the diameter and thickness of the disc, and how far are we from the centre?

The distant spiral galaxies, like the one in Andromeda, give us a model for our own. Most of the bright stars are in a flattened disc, arranged in more or less regular spiral patterns. The centre is a

This famous composite of the celestial sphere, with the plane of the Galaxy as equator, was made by the Lund Observatory in Sweden and illustrates graphically the concentration of stars in the Milky Way. The two bright areas to the south are the Large and Small Magellanic Clouds.

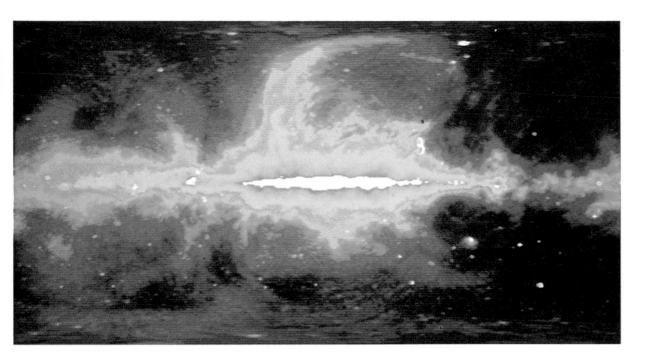

thickened boss to the wheel-like structure. Around the whole structure is a nearly spherical halo of fainter stars.

The flattened shape of spiral galaxies is due to their rotation. If we imagine a galaxy forming out of a vast cloud of primaeval gas, any slow rotation of the cloud will speed up as gravity shrinks the cloud. The cloud flattens, and the outer parts form a disc. Stars form, some with sufficient speed to escape from the disc and form a halo.

The large-scale structure of our Galaxy fits this model well, but near the Sun we are easily confused by local detail. The brightest visible stars do indeed form a belt across the sky, but this belt does not quite coincide with the average line of the Milky Way. This line of local bright stars is Gould's Belt; it runs through Orion and Taurus in the north and Lupus and Centaurus in the south. It is an accidental, minor collection of local stars, a temporary ripple in the major whirlpool. We can also see structure in the disc, which is discernible as spiral arms, but we are too deeply immersed in the disc itself for this to be obvious. Add the problem of the dust clouds, which actually obscure our view in some of the most interesting directions, such as in Serpens and Ophiuchus, and the advantages of the radio map become obvious.

We start fitting our Galaxy to the model by looking towards the poles rather than the equator, so avoiding the problem of the dust clouds. We measure the thickness of the disc from the distances of various visible objects, and it soon becomes obvious that there is a thin disc of bright stars and a much thicker halo of old stars.

Bright, young stars, of spectral types O, B and A, such as γ Velorum, Rigel, Sirius and Vega, are the most tightly concentrated in the plane of the Galaxy. They are mainly within 400 light years of the plane. The older stars, planetary nebulae such as the Ring in

This colour-coded map of 408-megahertz radio emission from the whole sky has a resolution of about 1° and was compiled from data collected over 15 years from a number of observatories. The most emission is concentrated in the plane of the Milky Way and shows as white. A number of well-known individual sources, both within the Galaxy and extragalactic, also stand out.

The relatively nearby Andromeda galaxy is two and a quarter million light years away. The giant spiral is accompanied by two small elliptical galaxies.

Lyra and red giants such as Betelgeuse are less concentrated, lying within 2000 light years of the plane. The young stars are still associated with their dusty origins near the galactic plane, while the older stars seem to have escaped.

Globular clusters, and some other old stars, are distributed entirely differently; they represent a distinct population which forms a nearly spherical halo round the centre of the Galaxy, and they are concentrated towards the centre. If we look towards the poles we see few globular clusters, but they are spread to very large distances. Globular clusters are best seen towards the centre of the Galaxy; half of them are found within 30° of the centre in Sagittarius, even though large parts of the sky in that region are obscured by dust clouds. The best examples of globular clusters are ω Centauri in the southern hemisphere and M13 in Hercules, both of which are seen to be detached from the Milky Way. Globular clusters and other old stars are found at distances up to 100 000 light years from the centre of our Galaxy.

The spiral arms

The radio map that gives the clearest picture of the Galaxy as a whole shows little of the spiral arms typical of disc-like galaxies. A different radio technique, observing the narrow-band radiation from hydrogen gas, gives a far more detailed picture.

The stars, and the gas, of the disc revolve round the galactic centre in nearly circular orbits. As in our planetary system, the orbital velocities are greatest near the centre, but the velocities vary

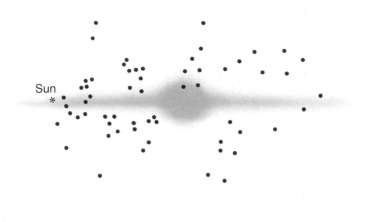

The distribution of globular clusters in a spherical halo around the Milky Way galaxy is shown here schematically.

Sun

100 000 light years

with distance from the centre of the system in a more complicated way. Given this relation, the radio emission from any region of hydrogen gas tells exactly where it comes from, as we can measure velocity by the Doppler shift of the radio frequency. The result is a beautiful delineation of a spiral arm structure, seen as clearly as it would be by an observer looking down on the galactic plane from another galaxy.

Given time, the same outside observer would see the whole disc rotating like a whirlpool, and the spiral arms would seem like temporary ripples. They are indeed temporary. The speed of rotation is much faster in the centre than on the outside, and if the arms were permanent they would be very tightly wound. The Sun takes about 250 million years to make one circuit round the Galaxy, so that our part of the Galaxy has completed about 40 rotations; the inner parts have made many more.

Evidently the arms are waves of temporary concentrations of gas and stars. The concentration is originally made up of gas and dust; the stars are there because concentrated gas and dust condense to form new, young stars. The little we can see of the spiral arms without using a radio telescope confirms this idea. The stars in spiral arms are young stars, and it is the young stars that show us where the arms run across the sky.

Look again at Orion. Here we see young stars, like Rigel and the stars in the Belt; beyond these stars we have the Orion Nebula and the Molecular Cloud. We are looking along a part of a spiral arm: as it happens, not a smooth well-organised arm, but along a spur attached to the main spiral. The stars in the Belt are only about 8 million years old, and the youngest stars in Orion are only about 2 million years old.

The marked contrast in age with the stars forming a galactic halo is crucial: globular clusters, for example, contain stars that are over 10 000 million years old. In this time, the spiral arm pattern can form and re-form many times over.

The Orion spur joins on to a spiral arm known as the Perseus arm. This is the major outer arm of the Galaxy, and we see it

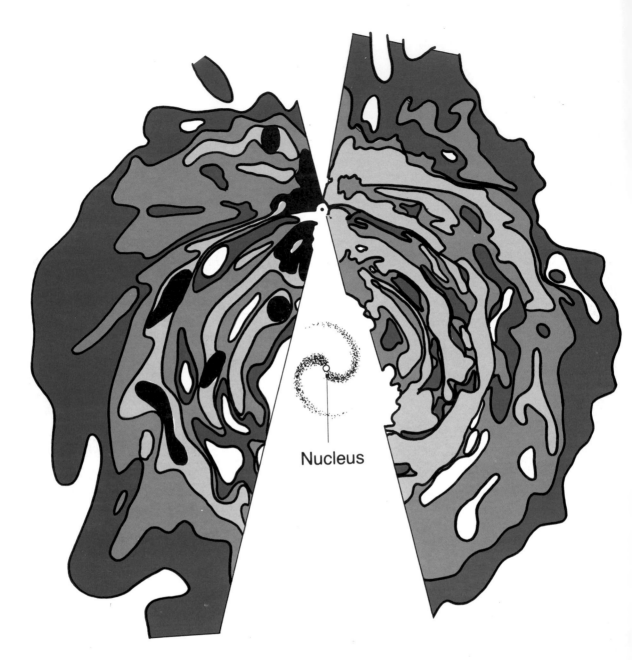

Nucleus

Radio emission at a wavelength of 21 centimetres generated by neutral hydrogen atoms in interstellar matter can be used to map the spiral arms of the Milky Way. The radiation from different arms can be distinguished because they are moving at different speeds. The Doppler effect causes the wavelength of the radiation to change by an amount that depends on the speed.

contours

stretching across the northern sky from Cygnus through Cassiopeia and Perseus to Taurus. The double cluster h and χ Persei is newly formed from this arm, and a dark nebula in Cassiopeia, known as W3, is another known birthplace for new stars. Like that in Orion, this nebula contains the simple molecules that are detected by the radio spectral lines.

The Galactic centre

Hidden behind the dust clouds of Sagittarius is the pivotal point of galactic rotation. As for many other galaxies, the centre is far more than a mere geometrical construction. Radio and infrared waves penetrate the dust, and the maps constructed by radio telescopes and satellite-born infrared telescopes show us an astonishing con-

Dense clouds of stars swarm in the direction of the centre of the Galaxy, which lies in the constellation Sagittarius. The Galactic centre itself is concealed by opaque dust clouds.

centration of activity. At the centre of this activity is the most massive object in the whole of the Galaxy.

The Sun is about 30 000 light years from the Galactic centre. Outside a radius of 10 000 light years the patterns of spiral arms seem to be moving in circular orbits round the centre. Inside this radius the pattern is more complicated. At 10 000 light years there is an arm which is expanding as well as rotating. Closer to the centre there is another pattern of arms which is tilted a few degrees out of the plane of the Galaxy. It seems that something near the centre disturbs the even pattern of rotation.

The central point of radio maps of Sagittarius, which contains the centre, is the object Sag A. This is a hot mass of gas, elongated along the galactic plane. Within it lies Sag A west, and within that again lies the infrared source IRS 16. This appears to be the true nucleus of our Galaxy; it could, however, be a foreground object, accidentally on our line of sight to a darker, true nucleus.

Around the nucleus there is a ring of hot gas, with a radius of about 5 light years. Outside the ring there is a cloud of gas and dust, like the more familiar Giant Molecular Cloud in Orion. At the inside edge one of the simplest molecules, hydrogen, is heated to 2000 K by a violent wind blowing outwards from the nucleus.

There are several clues to the nature of the nucleus. First, assuming IRS 16 is indeed the nucleus, the total amount of infrared radiation requires an object ten million times the luminosity of the Sun. Second, the gas clouds in the ring are orbiting round the centre at a speed which requires a concentration of about ten million solar masses to hold them in place. Third, the rate of mass loss in the violent wind is large; it corresponds to at least one million times the normal rate for a normal hot star. So, the nucleus seems to be a very massive object. Is it a black hole, a single very massive star, or a tight concentration of normal stars?

The observation of the violent outward wind, if it is correct, tells us that the nucleus cannot be a black hole. Black holes attract an infalling wind: they can suck but not blow. Beyond that, we cannot be sure. But we know that within a volume that in our part of the Galaxy contains on average only one star, there is a mass equivalent to ten million times that of the Sun. When we look at other galaxies, and particularly the radio galaxies, we will find that even this fantastic concentration is greatly exceeded.

A manifestation of the power of this central engine is demonstrated by the radio emission from the central 500 light years of the Galaxy, showing long filamentary arcs crossing the plane. These streaks of radio emission are about 130 light years long; each strand in the tangle of threads is only 3 light years wide. The filaments are almost certainly tracing out a strong magnetic field, which is connected in some way to the nucleus. Here is another familiar astrophysical process. As in pulsars, the central engine is rotating rapidly and has a massive magnetic field; it is a dynamo. When we know enough about the Galactic centre to construct a proper picture of it, we will be able to show how the gravitational energy of collapse has been transmuted into electrical energy by a dynamo. The galactic centre deserves its nickname of the central engine of our Galaxy.

18

The galaxies

Anyone interested in the night sky is likely to recognise the words Andromeda Nebula, and may also know that this represents one of the nearest galaxies, about 2 million light years away, and that there may be as many galaxies beyond it as there are stars in the Milky Way. The Andromeda Nebula appears as a faint patch of light, best found by tracing along the constellation of Andromeda from the square of Pegasus.

Observers in the southern hemisphere have an easier start to the exploration of the depths of the Universe. They can see the two conspicuous clouds, looking like part of the Milky Way but clearly separate from it, which are our nearest neighbour galaxies. These are the Magellanic Clouds, named after the explorer who circumnavigated the globe in 1518–20. To the naked eye, there is no indication that these are whole galaxies of stars, detached from the Milky Way and an order of magnitude further away than the centre

How to find the Andromeda galaxy. It actually extends over more than two degrees of sky. The bright central bulge can be seen with the naked eye as a faint misty patch near to the star ν Andromeda.

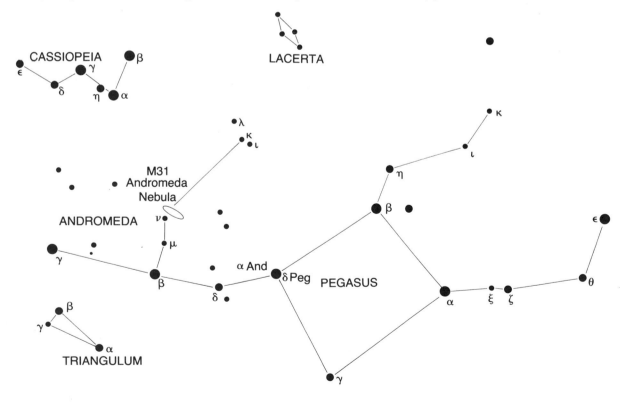

The Large Magellanic Cloud, 160,000 light years away, is the galaxy nearest to the Milky Way and a member of the Local Group of galaxies. The huge Tarantula Nebula, a glowing gas cloud within the LMC, is a striking feature of the galaxy. On this photograph taken in May 1987, the supernova can be seen close to the Tarantula Nebula and Comet Wilson is to the left of the galaxy.

of our own Galaxy. Binoculars or a small telescope will allow us to explore a little further, but a more interesting view needs a telescope with an aperture of at least 20 cm. For example, the brightest members of the Virgo cluster of galaxies may be seen, located on a line between the stars ε Virginis and β Leonis. Even so, the sight of these more distant galaxies is somewhat disappointing: they are no more than faint smudges of light, and their true nature is revealed only by photography.

The first catalogue of nebulae was made by Messier in 1784. He was actually searching for comets, and was concerned only to record the positions of fixed nebulous objects which might be wrongly reported as comets. His catalogue starts with M1, the Crab Nebula, which is of course a supernova remnant; only one-third of the 104 objects listed by Messier are extragalactic. The Andromeda Nebula, for example, is M31 in this catalogue. The catalogue of nebulae most used today, the New General Catalogue (NGC) was originally compiled by William Herschel and his son John, who provided the southern hemisphere survey. Its present form is due to John Dreyer of Armagh Observatory in Northern Ireland. As with Messier's catalogue it contains objects in our Galaxy as well as the extragalactic nebulae.

The beautiful spiral form of many of these distant galaxies was first revealed to Lord Rosse, when he started to use his 6 ft diameter telescope in 1845. His sketch of M51 is reasonably close to reality, but Herschel's 2 ft diameter telescope, 60 years earlier, could only give a very sketchy outline. In the 1920s photography and the

The 'Whirlpool' spiral galaxy (M51) in the constellation Canes Venatici is a pair of interacting galaxies about 35 million light years away.

The spiral structure of M51 was first discovered by Lord Rosse when he started to use his 6-foot telescope in 1845 and is clearly shown in the sketch he made.

100 inch telescope at last enabled Hubble to survey several hundred galaxies, and to show us their remarkable range of forms. Even more remarkable are their distances and their velocities, revealed by the redshifts in their spectra.

Types of galaxies

Galaxies are often described as either elliptical or spiral, with a third category of irregular for the misfits. In fact, all categories display a mixture of order and disorder, and there is such a variety that experienced observers can recognise and name photographs of hundreds of individuals. Why is there such a variety? Can we see in it a grand design, possibly a hierarchy of sizes or ages? Does it contain clues to the evolution of the Universe?

Now that we know that the whole Universe is active and changing, we can more easily comprehend the dynamics and evolution of the galaxies. The ellipticals, which include the most massive of all

The peculiar galaxy NGC 5128, known to radio astronomers as Centaurus A because it is the brightest source of radio emission in that constellation, is a giant elliptical galaxy straddled by dark dust lanes.

galaxies, seem to be the most stable; if there is evolution from one type to another, these must be the end product. M87 is one of the largest, containing 10^{13} solar masses. There is a suspicion that very massive galaxies like M87 can swallow up lesser ones, sucking them in by the force of gravity: this may be happening in the galaxy Cen A, where the central part is obscured by a huge in-falling dust cloud, which looks like a spiral galaxy seen edge-on.

The spiral galaxies themselves are more obviously evolving, both in shape and in their population of stars. The flat disc of a rotating spiral is rotating, but not at a uniform angular velocity. The arms may therefore be expected to trail behind the central parts, and wind up very tightly in the lifetime of a galaxy. In fact they remain open and they are therefore waves of condensation rather than permanent groups of stars. It is the gas and dust of the galactic plane which condenses and relaxes in the wave; in the condensation new stars are born which delineate the spiral arms. M33, a nearby galaxy, shows bright patches along the arms where new stars have recently formed. These new, hot stars are very massive; they evolve rapidly and are no longer conspicuous when the density wave has moved on.

The stars in an elliptical galaxy are also in orbit, but their orbits are at random inclinations rather than confined to a plane, as in the spirals. It is in fact the gas and dust which form the plane in a spiral

The 'Pinwheel' galaxy (M33) in Triangulum is a member of the Local Group and one of the spiral galaxies nearest to the Milky Way at a distance of 2.4 million light years.

The 'Sombrero' galaxy (M104), 44 million light years away. The huge nuclear bulge contains many old red stars. The white colour at the centre is due to over-exposure. The blue light comes from hot young stars and dust clouds cause the dark band in front of the nucleus.

galaxy. The stars form out of condensations in the dust, showing up the plane and its spiral pattern of condensation. The Sombrero Hat Galaxy, seen edge-on, shows the dust plane as a dark lane. Dust cannot move in random orbits: if it does it collides, and all motion is eventually lost except in organised orbits in a well-defined plane following the galactic rotation.

Spiral galaxies, like the Milky Way, have two distinct stellar populations: the old and the young. The old stars are in a nearly spherical halo, like an elliptical galaxy; the young are in the spiral arms, forming the flat disc. The irregular galaxies, which are usually the least massive, also usually have the highest proportion of dust and gas; they may contain none of the old, halo population of stars (technically known as Population II). The smallest galaxies, naturally known as the dwarfs, may contain fewer than a million stars.

The local group and other clusters

Although the most distant galaxies are uniformly distributed over the sky, on a smaller more local scale their distribution in space is very non-uniform. On the smallest scale they form distinct clusters, one of which includes our own Galaxy, the Milky Way. This local group also contains two other spirals (Andromeda and M33), two irregulars (the Magellanic Clouds) and a collection of two dozen or more minor members. Most of these minor members are irregular in shape, but two dwarf ellipticals can be seen close to the Andromeda Nebula. The local group also contains another large galaxy, Maffei 1, which is almost hidden by the Milky Way. This

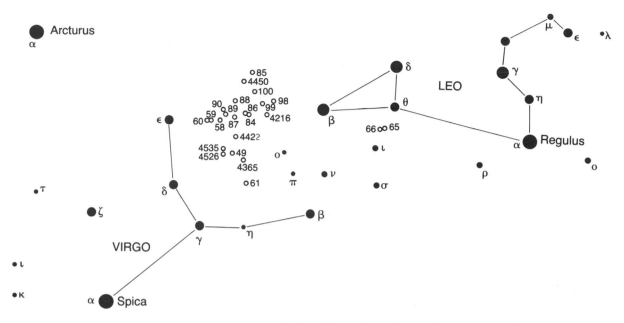

The small open circles on this chart mark some of the brighter galaxies in the relatively near-by Virgo cluster, which covers an area of sky 10° by 12°. The huge Virgo cluster is about 60 million light years away and contains hundreds of galaxies.

NGC 6822 is a small irregular galaxy in the Local Group. It is about two million light years away and ten thousand light years across.

A cluster of galaxies in the constellation Centaurus.

galaxy is best seen by infrared telescopes; it is either elliptical, or in an intermediate configuration between elliptical and spiral known as lenticular.

The Virgo cluster is much larger, with at least 1000 members. As with most of the larger clusters, it is more orderly than the local cluster both in shape and composition. It contains a high proportion of ellipticals, including the massive M87. Large clusters have a strong gravitational pull, and it may be that they can swallow small clusters which cross their paths. Only the general expansion of the Universe prevents this happening to most clusters.

Clusters of galaxies, like those in Virgo or in Coma Berenices, stretch across many degrees of the sky. More distant clusters, like that in Leo show up as compact collections.

Our two-dimensional view of the sky suggests that galaxies are fairly uniformly spread over the sky, with clusters of various sizes superposed on a background population. This is an illusion. Using redshift as a measure of distance (see p.211), the whole three-dimensional pattern is revealed.

Two unexpected features stand out. First, there is no background population: instead there are vast empty spaces between the clusters. One such empty void is 40 million light years across. Second, the clusters are not neat spheres: instead they are like filaments in a tangled net. We may not be right to regard even the giant Coma and Virgo clusters as isolated units, dynamically stable like the globular star clusters in our Galaxy.

Is there any larger-scale order in this chaotic tangle of clusters? It is often suggested that there is a larger unit, the supercluster. At this stage, however, it is difficult to prove that an apparent large-scale collection of clusters is truly a physical entity rather than a random fluctuation in the population. In our discussion of cosmology (Chapter 20) we regard the clusters as the basic building bricks of the present-day Universe.

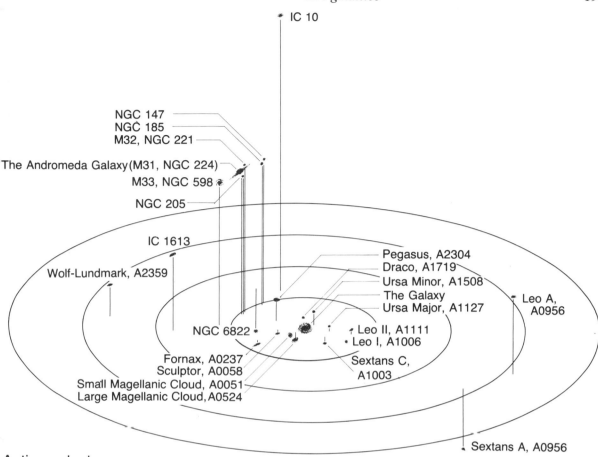

The distribution in space of the Local Group of galaxies. The distances between the galaxies are small in relation to the sizes of the galaxies, the distance to the nearest large galaxy being only twenty galactic diameters.

Active galaxies

The apparent calm symmetry of the giant elliptical galaxy M87 hides a remarkable manifestation of tremendous activity in its nucleus. A photograph made with a short exposure so as to show only the bright nucleus, reveals also a bright jet of light emerging from it. The same jet shows on a map of radio emission from this galaxy; here the stars do not radiate at radio wavelengths, but the jet is a prominent feature, extending outside the visible galaxy. Radiation from this jet is totally different from starlight. It is called synchrotron radiation, by analogy with radiation observed from electrons accelerated to very high energies in the accelerator machines of laboratory particle physics.

Synchrotron radiation comes from electrons with very high relativistic energies moving in a magnetic field. The total energy in this jet of electrons is equivalent to the mass of some hundreds of stars, all transformed into energy according to Einstein's law that the energy equivalent to a mass m is mc^2, where c is the velocity of light. The jet seems to emanate from a central nucleus; but this nucleus must be a powerhouse of a totally different order of magnitude compared with the nuclear furnace of a star.

Not many galaxies show direct visible signs of such central energy as does M87, but there are many other manifestations of activity which indicate that galaxies often have very remarkable centres. Even our own Milky Way has a concentration of matter at

The giant elliptical galaxy, M87, at the heart of the Virgo cluster of galaxies, is probably very nearly spherical. Its total mass is estimated at three hundred billion Suns, making it the most massive galaxy known. This exposure has been made to bring out the halo of at least eight hundred globular clusters, each containing hundreds of thousands of stars. The halo is about 120 000 light years across.

A short exposure photograph of the galaxy M87 reveals the presence of a peculiar bright jet, originating in the nucleus and consisting of an aligned series of six 'knots'.

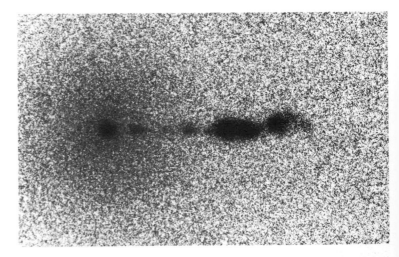

its centre, with a mass of ten million suns, and there is powerful radio emission from regions close to it. A class of active galaxies, the Seyferts (named after Carl Seyfert), have nuclei which emit strong X-rays and radio waves. Close to the nucleus of one of these, NGC 4151, there are gas clouds in orbit at 1500 km s^{-1}; at this high speed they must be in orbit round a very massive nucleus, with about 100 million solar masses. Cen A, the elliptical galaxy whose centre is obscured by a band of dust, also has a small and very active nucleus which can be seen both in infrared and X-ray maps. The

X-ray emission is variable, which again indicates that the nucleus is very energetic and very small.

Radio galaxies, the subject of the next chapter, all derive their properties from an unseen nuclear powerhouse. In the much rarer but more spectacular quasars the nucleus itself outshines the rest of the galaxy by a factor of a hundred or more.

Activity in galaxies is not, however, confined to the nucleus. M82 was for a time called the Exploding Galaxy, but its extra-ordinary,

High resolution radio maps of M87 show that the radio emission comes from clouds on either side, one of which is fed by a prominent jet. The colours indicate the strength of the radio emission.

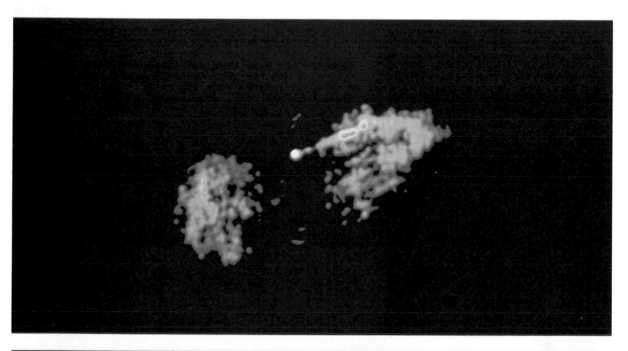

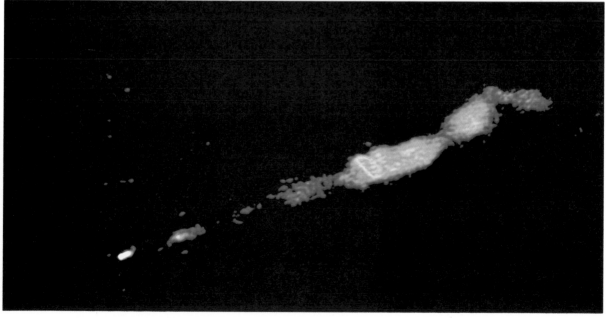

This peculiar galaxy, called Arp 243, was found by the IRAS satellite to be one of the most luminous galaxies in the infrared region. More than 95 per cent of its energy is emitted in the infrared. It may possibly be two giant spirals that have collided and merged. Astronomers think that the interstellar material has suffered a shock that caused it to collapse and produce new stars at a rate one hundred times greater than occurs in normal spirals. The stars are buried in dust that absorbs energy from the starlight and re-emits it as infrared radiation.

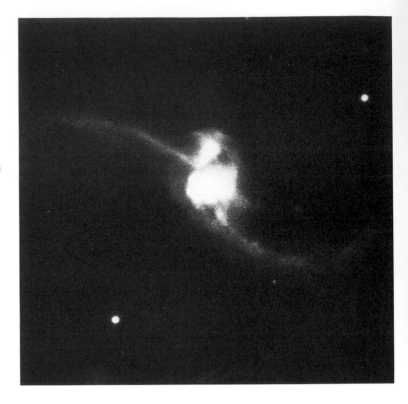

ragged appearance is now believed to show the opposite; dust and gas is falling into the galaxy under its strong gravitational pull. The suspicion is that this galaxy has collided with a dark cloud of dust, possibly like one of the faintest irregular galaxies of our local group.

Dust clouds in the Milky Way, such as the Orion Nebula, are the birthplaces of stars. The same must occur in galaxies such as M82, but it is not obvious from ordinary photographs. Infrared radiation, which shows up cooler objects than visible light, is a good indicator of dust clouds. The satellite infrared telescope IRAS surveyed the whole sky for galaxies with unusually strong infrared emission; M82 was an obvious example. Another was NGC 253, a spiral in the Sculptor Group. This is one of the brightest infrared emitters, and the indications are that it is currently producing unusually large numbers of new stars from its thick dust cloud. Galaxies such as M82 and NGC 253, where this is happening, are called starburst galaxies. A good example is the galaxy known as Arp 243, found by IRAS to emit 95% of its energy in infrared radiation. This all comes from dust clouds warmed by the ultraviolet light from newly formed stars.

Collision and mergers

Most galaxies are at first sight independent units, interacting with one another only by long range gravitational attraction. This is an illusion due to the invisibility in ordinary telescopes of the gas which exists between the galaxies. A local example is the stream of gas pulled out of the Magellanic Clouds by our Galaxy, forming an arc over the sky which is detected by its hydrogen line emission. A more substantial gas cloud seems to exist in all large clusters of galaxies: in this case it is detected by X-ray telescopes, which show that every cluster has more X-ray emission than the sum of the radiation from its individual galaxies. Furthermore, this cluster emission comes from hot gas, at 10^8 K, and it even includes the spectral line radiation with photon energy 7000 electron volts from highly ionised iron atoms. The mass of this intracluster gas forms a substantial fraction of the total mass of the cluster. Its high temperature suggests that, unlike the Magellanic Stream, it is falling into the cluster rather than emanating from one of the galaxies within the cluster.

There is, therefore, a growing suspicion that substantial quantities of gas and dust are encountered by galaxies, and by clusters of galaxies, in their random movements within the frame of the expanding Universe. The in-fall of gas into M82 may be the result of the collision of a normal galaxy with a gas cloud. In-fall into a galaxy with a massive nucleus may produce the more dramatic active galaxies, such as M87 with its jet of synchrotron radiation. This may be the explanation of Cen A, in which the dark dust lane may be a spiral galaxy, full of dust and gas, falling into a more massive elliptical. The massive nucleus collects the in-falling material, which may already have formed stars, and the gravitational energy of the fall is released as X-rays and other radiation.

Collisions of many kinds certainly do occur. Sometimes they are very dramatic in appearance, especially when spiral galaxies are involved. It is hard to imagine that these tortured systems will ever settle down to a single symmetrical galaxy.

The two galaxies NGC 4038 and 4039 (known as 'the antennae') are so close to each other that their mutual gravitational attraction has distorted their structure, producing long streamers of stars and gas. They are about fifty million light years away.

We do not yet know whether the radio galaxies and quasars are excited by accretion of gas, although it seems that their energy comes from gravitational forces close to a large central mass. It is a reasonable theory that they represent the extreme of a continuous range of activity in galaxies; the active galaxies described in this chapter are then the nearby and well-studied prototypes for the rarer, more distant and less well understood subjects of the next chapter.

Distances and velocities

The simple statement 'our Universe is expanding' depends on a remarkably simple observational fact: the further away the galaxy, the faster it moves away from us. How do we measure the enormous velocities and distances involved in this cosmic expansion?

Velocity is the easier quantity to measure, using the redshift of features in the visible spectrum. Many active galaxies emit bright spectral lines from their nuclei; these can give a direct measurement of redshift, which is related simply to velocity by the Doppler effect (see Chapter 19). The light from most galaxies, however, is the sum of starlight from all types of star; consequently it has no obvious sharp spectral lines. Nevertheless there may be a step in the continuous spectrum, due to neutral hydrogen, whose position depends on velocity. Again, the radio spectral line from hydrogen, at 21 cm wavelength, gives very accurate velocity measurements for the brighter galaxies.

The distances of the nearest stars can be measured geometrically by observing their apparent change in position as the Earth moves in its orbit round the Sun. The next stage in determining distances is to compare the brightness of one of these nearby stars with the brightness of a distant one of the same type; if it can be assumed that the intrinsic luminosity of the two is the same, the ratio of their distances can be found from the inverse square law. For example, five magnitudes difference corresponds to a ratio of 100 in brightness and a ratio of 10 in distance, provided that the stars have the same intrinsic luminosity, or absolute magnitude.

The problem in this extension of the distance scale is to find identical types of star, whose absolute magnitudes can be relied on to be identical. The Magellanic Clouds provide an essential link in these distance measurements. They are near enough for individual stars to be seen: the brightest reach ninth magnitude. Among the stars of the Small Magellanic Cloud are a collection of periodic variables of a particular type, the Cepheids, whose absolute magnitude is related precisely to their period. Cepheid variables can be observed in the Andromeda galaxy, and the ratio of distances between these two galaxies can be found from the ratio of apparent brightness.

A series of links, each with an element of uncertainty and error, is needed to connect the distances of the most remote galaxies to our laboratory standards of length. The relation between redshift and distance is consequently known only to about 30%. Cosmologists do not worry too much about this; they merely use redshift to infer the great distances involved. Radio galaxies and quasars have very large redshifts, as we will see in the next chapter; we rely totally on these for our measurement of their distances.

19

Radio galaxies and quasars

A few decades ago, observational astronomy involved only the analysis of light received on Earth from various celestial bodies. The idea that other kinds of radiation, radio, infrared, ultraviolet, X-rays and gamma-rays, could also be transmitted, received and analysed would have been hard to grasp when radio meant only broadcasting, X-rays meant medical examination, ultraviolet meant only suntan. These are, however, all electromagnetic waves; they can reveal to us a hitherto concealed Universe, comprising some strange objects which previously were unremarkable or even invisible. Such is the story of the radio galaxies and the quasars.

Radio waves and light are often emitted by the same object, but they reveal different aspects of it. The Sun, for example, emits light from the familiar yellowish sphere called the photosphere; it emits radio waves from a much larger and hotter gaseous region called the corona (see Chapter 12). The Sun was the first individual celestial object found to emit radio waves (that is excluding the emission from the Milky Way discovered by Jansky in 1932): these days we might say it was the first object to be 'seen' by a radio telescope.

There were other radio sources whose radiation could be picked up by radio telescopes; these were outside the solar system, and could be located against the background of stars and galaxies. In 1952 one of them, in the constellation of Cygnus, was located sufficiently accurately (by FGS working at Cambridge) for a special search to be made for any visible counterpart. This search was a turning point in modern astronomy. Walter Baade and Rudolf Minkowski, working with the large optical telescopes in California, found that the radio waves were coming from a faint (16th magnitude) galaxy of unusual shape, with an unusual spectrum, and with a large redshift. Measured as the ratio of the shift in wavelength $(\triangle\lambda)$ to the laboratory wavelength (λ), the redshift $(z = \triangle\lambda/\lambda)$ was 0.057. The cause of the redshifts seen in the spectra of galaxies is generally thought to be their motion away from us, the speed of which increases with distance. So this discovery demonstrated that radio astronomy could be used for penetrating to great distances: the radio source Cyg A was easily detected, and a radio telescope might be able to detect a similar radio source up to 100 times further away.

In 1960 another radio source, 3C 295 in the third Cambridge catalogue, was identified with a visible galaxy. The redshift of this galaxy was found to be $z = 0.45$. Many radio sources had been discovered by that time, but few had been identified. From several

The radio source Cygnus A is one of the brightest objects in the sky at radio wavelengths yet its optical counterpart is a peculiar galaxy of only 16th magnitude.

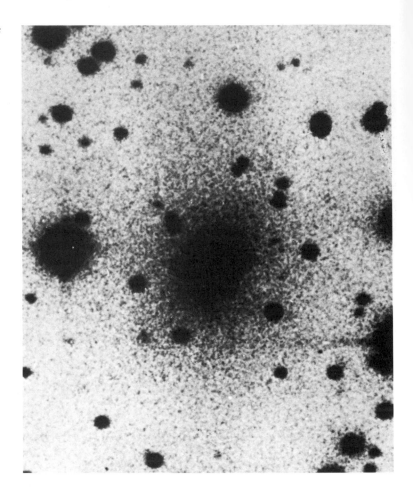

of them the radio waves appeared to come from a pinpoint of a source rather than the more diffuse area of the radio galaxies. The problem was resolved in 1963, after the positions of three of these pin-point sources had been found to coincide with some unusual star-like objects. They were at first thought to be stars in our Galaxy. The spectra of these 'stars' included bright emission lines typical of hydrogen and magnesium, but at large redshifts. The first to be identified, 3C 273, had a redshift of 0.158 and it was soon found that the redshift of the other 'blue stars' was even larger. Either they were extragalactic, or the redshifts arose from some other effect not associated with the expansion of the Universe. These point-like radio sources soon became known as 'quasi-stellar' objects, or quasars. If they were extragalactic, as expected from the large redshift, they were probably bright galactic nuclei, and there should be a visible galaxy attached to them. At first no such galaxy could be seen around any of the quasars, but several years later such galaxies were indeed found – very faint objects whose nuclei had become hundreds or thousands of times brighter than the whole galaxy. Until they were seen, there was some doubt about the meaning of the redshift. We now know that the quasars are indeed at very great distances; the most distant known has a

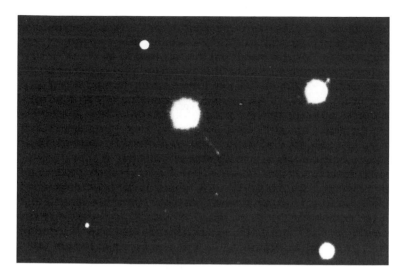

This optical image of the quasar 3C 273 reveals a faint luminous jet. If the quasar is at the distance suggested by its redshift, the jet is 150 thousand light years long and its infrared power output exceeds that of the entire Milky Way by a factor of at least a hundred thousand.

redshift of 4.4, where it must be receding from us with a velocity greater than 96% of the velocity of light. These distant quasars have two messages for us: they represent some remarkable physical processes, and they provide our most penetrating data in cosmology.

Mapping the radio galaxies

The radio telescopes used in locating the radio galaxies and quasars were interferometers, using two separate antennae connected to the same radio receiver. The spacing between the pair of antennae is important: an interferometer with a large spacing will only respond to radio sources with a small angular diameter. Michelson adopted the same principle when he used an optical interferometer with variable spacing to measure the angular diameter of some red giant stars, including Betelgeuse. The same technique, applied to the radio galaxy Cyg A, showed that it was several arc minutes across rather than a few arc seconds, as seen on the optical photographs.

The large overall angular extent of the radio galaxies, and the detailed structure within that total extent, can be seen in many maps of their radio emission. These maps are now made with astonishing angular detail; resolving powers of 1/10 arc sec are commonly used, far better than the resolution of any optical telescope. The angular resolving power of a telescope is, however, limited to the ratio λ/d, wavelength divided by the diameter of the telescope aperture. How can radio, with wavelengths of centimetres or metres, produce better resolution than light, with wavelengths as short as half a micron? The solution is simple: as explained in Chapter 4, interferometers are used with very large spacing d. The Jodrell Bank MERLIN array of telescopes, for example, uses spacings up to 200 km.

Interferometers give more than a measurement of angular diameter. Each pair of telescopes in a multi-telescope array gives an interferometer recording of a particular component of the distribution of radio brightness across the radio source; mathematically

The location of the radio telescopes used in the Jodrell Bank Multi-Element Radio-Linked Interferometer Network – MERLIN for short.

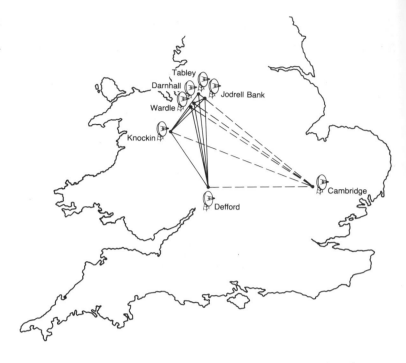

this is known as a Fourier component of the brightness distribution. The many different interferometer pairs each record a set of Fourier components over an observing period of 12 h, during which time the rotation of the Earth ensures that the Fourier components cut across the source over as wide a range of angles as possible.

Reconstructing the radio brightness distribution from the Fourier components recorded by a network such as MERLIN requires a large computer and some special image-processing techniques.

A larger resolving power is needed for some radio galaxies. Fortunately larger spacings are available in a European network of radio telescopes. Maps are now often made by combining MERLIN with this European network. The recording system, which uses videotape recorders, is known as Very Long Baseline Interferometry (VLBI).

Mapping the quasars

Although the angular diameters of many of the radio sources were fairly easily measured with simple interferometers, some were remarkably resistant. Larger and larger spacings were used, so that the interferometers had to be connected by radio links rather than by cable. A memorable step was made in 1956 by Henry Palmer, who took a portable antenna to the Cat and Fiddle Inn on the top of the Pennine Hills, 20 km from Jodrell Bank, and set up a radio link to a fixed antenna at Jodrell Bank. Several radio sources remained unresolved so in the succeeding years the portable antenna was moved further and further away from Jodrell Bank. In 1960 it was near Holywell in North Wales but seven out of nearly 100 sources measured were still unresolved – indicating that their angular

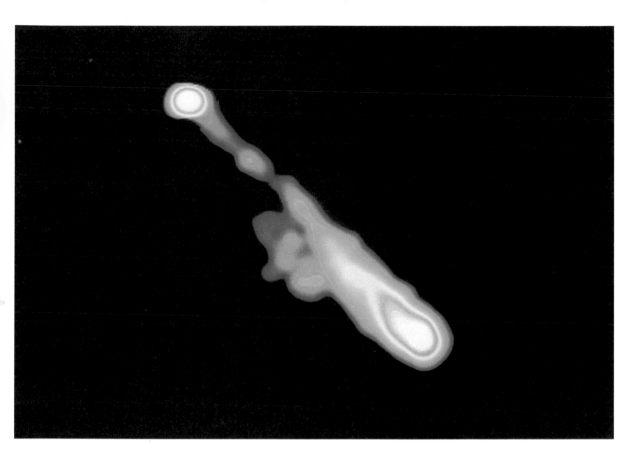

diameters must be less than 3 arc sec. It was these measurements which stimulated the optical search that led to the discovery of the star-like objects in 1960 and then to the discovery that they were distant extragalactic objects – the quasars as already related.

Palmer and his group at Jodrell Bank continued their efforts to increase the resolving power of the interferometer, and in 1965 the telescope at Jodrell Bank was linked to one near Malvern. A number of sources were still unresolved, indicating that their angular diameters were less than 1/10 arc sec.

Their diameters remained unresolved until the interferometer baselines were increased, using VLBI, to thousands of kilometres by using an international network of radio telescopes. On this very large scale, and by using short wavelengths so that λ/d is as small as possible, it became possible to map even these pin-points of radio emission, the quasars. In many cases, however, the maps are rudimentary, showing some sort of elongated object characterised only by a length and a diameter, barely resolved at the maximum interferometer spacings available on Earth. (There are plans to put a radio telescope into a satellite orbit, so as to obtain even larger interferometer spacings.)

Maps of the most distant objects in the Universe are not expected to show changes from year to year. In a series of maps of the quasar 3C 273, at intervals of around one year, surprisingly, the quasar appears to be double, with a spacing that grows rapidly from one

A radio map of the quasar 3C 273, showing the core of the quasar (top left) and the strongly radiating jet. This map was made with the MERLIN radio telescope.

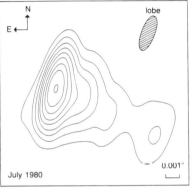

A series of detailed maps of the core of
3C 273 showing the ejection of a cloud
of radio-emitting material. The apparent
speed of ejection is greater than the
velocity of light; the actual speed has
been magnified by a perspective effect.

map to the next. At the large distance of this quasar, the two components appear to be flying apart at several times the velocity of light! An actual 'superluminal' velocity is impossible, and this rapid separation is now attributed to a perspective effect: the separation of the components must be nearly along the line of sight to the quasar, in which case the geometry of special relativity predicts that the angular separation will increase at such a rate. Nevertheless, the observation shows that the quasar has ejected a very energetic radio emitter which moves at a speed approaching that of light.

We can now see a sequence connecting quasars, radio galaxies, and the active nuclei of more normal galaxies. There is a central engine, or powerhouse, which in the quasars dominates the appearance of the galaxy. In the radio galaxies the jets of energetic electrons ejected by the quasar-like nucleus form clouds, which then become the dominant source of radio waves. Less energetic nuclei reach neither of these levels, but are still seen to be active, even if they are detectable only by their X-ray or infrared radiation. There are intermediate types, showing some of the characteristics both of quasars and of radio galaxies. A composite radio map of the galaxy NGC 6251, made using the complete range of angular resolution available from international VLBI through to the 5 km synthesis array at Cambridge, shows that the quasar-like nucleus is ejecting a tiny jet in a north-west direction. The jet continues in a turbulent but more-or-less straight line for a distance of half a million light years (more than the diameter of a normal galaxy); and the jet ends in a complex cloud, typical of one half of a radio galaxy. A similar cloud to the south-east may be at the end of an invisible jet in the opposite direction.

The structure of radio galaxies

The radio galaxy Cyg A is so powerful, and comparatively nearby, that it has become the archetype and the best mapped of all radio galaxies. A map made by the VLA radio telescope shows a thin jet leaving the nucleus and entering the cloud-like 'radio lobe' on the right. At the outside of this cloud is a flattened, bright region, where the jet is brought to a halt as it impinges on the diffuse intergalactic gas. The cloud then expands and turns back to envelop the jet.

On the left the jet is scarcely visible, but a very similar cloud shows that the central source emits symmetrical jets in both directions.

The clouds now stream back towards the nucleus, losing energy as they do so and emitting lower frequency radio waves. This low frequency radio emission shows up on a map made with the MERLIN telescope network. The clouds are seen extending back all the way to the nucleus, turning off to each side as they approach it. Similar characteristics can be discerned in many radio maps.

The designers of fusion reactors are interested enough in the physics of such well-confined jets and clouds of plasma; we, and they, are even more concerned to know what is the nature of the central engine which drives the quasars and the radio galaxies. The energy contained in the jet and the clouds is estimated at 10^{52} joules, corresponding to the energy of annihilating one million solar masses. Where does this energy come from, and how is it organised in this astonishing powerhouse?

Gravitation and black holes

The quasars are so far away, and their nuclei are so bright, that it is very difficult to study the galaxies in which they lie. There are, however, some rather commoner galaxies which show some of the characteristics of quasars, and which we can study in more detail because they are nearer. These are the energetic galaxies known as Seyferts. An example, NGC 1275, is found in the Perseus cluster; it is a bright radio galaxy, with radio clouds which are distorted in some sort of streaming motion, probably because it is moving through an intergalactic gas cloud. In another Seyfert, NGC 4151, the nucleus is surrounded by gas clouds whose visible spectral lines show orbital velocities up to 1500 km s^{-1}, higher than those of the gas clouds near the centre of the Milky Way. The nucleus must be more massive, probably nearer to 100 million solar masses. It must also be very compact.

The force of gravity is clearly the driving force in all these compact objects. A star falling into the massive nucleus of a galaxy can release more energy through its gravitational fall than it could if the whole of its mass was involved in a nuclear explosion. The more compact the central mass, the more energy is released. We cannot observe the nucleus directly, possibly for the fundamental reason that it is a black hole.

A radio map of the jet of NGC 6251, made with the Very Large Array. This jet is typical of the narrow, gently curved features linking the active nuclei of radio galaxies to their extended lobes of radio emission.

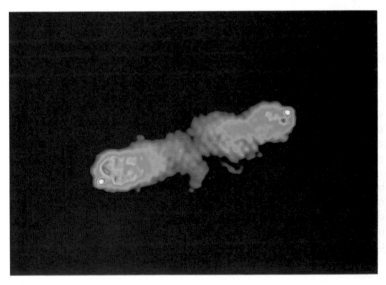

The detailed structure of the long-wavelength radio emission from Cygnus A shows up on this map made with the MERLIN network.

When stars collect together through their mutual gravitational attraction, they are usually kept apart, like the stars in our galaxy, by the revolution of the whole system about the nucleus. If there were no angular momentum, the system would totally collapse.

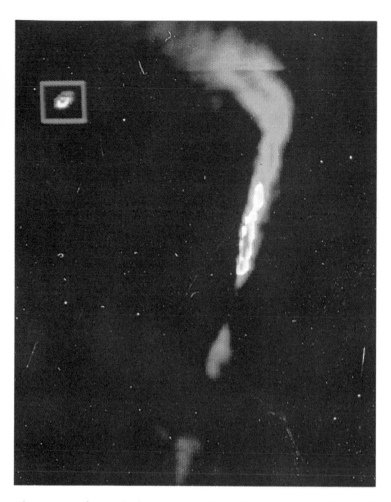

The radio jet of 3C 31, identified with the galaxy NGC 383, mapped by the Very Large Array.

There is no force that can resist the collapse of more than a few hundred solar masses, and the collapsing nucleus can become smaller than our solar system. The force of gravity at its surface then becomes so large that nothing, not even light or radio waves, can leave it. Such objects have been termed black holes, whose only external manifestation is the strong gravitational field.

Many astronomers now consider that it is reasonable to assert, but hard to prove, that black holes containing millions of solar masses exist at the centres of many galaxies. The active galaxies are those that are feeding the black hole with more stars; the radio and light signals that we receive from these quasars are the last signals that will be received from these stars as they disappear from view.

When Cyg A was identified as the first visible radio galaxy, its spectrum and its double nucleus were interpreted as a pair of colliding galaxies. The theory of cannibalism outlined above now suggests that this may still be the correct interpretation.

BL Lacertae – the variable quasar

Variable stars are common, and there are large catalogues listing them under the constellations in which they occur. BL Lac was long thought to be an irregular variable star with a featureless spectrum. Another similar object is AP Librae. These objects, and many others like them, are not stars at all, but another manifestation of quasar-like galaxies. They can often be detected as sources of radio, infrared and X-ray emission, and some are variable on time scales of a few days or even less.

BL Lac objects are rarer than quasars, but they are generally classed as a form of quasar in which the typical emission lines have been lost, possibly by broadening so that they merge with the continuum spectrum. A closer look at their spectra often shows instead a remarkable series of narrow absorption lines, which are also seen in distant quasars.

Absorption lines – intervening galaxies

High red-shift quasars and BL Lac objects often show narrow absorption lines at the long wavelength end of the spectrum, principally at wavelengths longer than a line due to hydrogen known as Lyman-alpha, which is commonly seen in emission. These lines occur in groups, allowing them to be identified as due to some abundant elements. These groups show a very large redshift in wavelength, although not usually as large as the redshift of the quasar itself. Each group of lines represents absorption in another galaxy, which happens to be on the line of sight to the quasar or BL Lac object.

A good example is provided by the BL Lac object 0215+015. (It has no other name, only these numbers which give its position in the sky.) It is variable, brightening from magnitude 19.5 in 1969 to 16.5 in 1977, and occasionally flaring to magnitude 14.5. This is bright enough for high resolution spectroscopy, which shows redshifted systems of absorption lines at $z = 1.345$, $z = 1.549$ and $z = 1.649$. Even within these groups there is fine structure: the lines with $z = 1.549$ show at least seven different components.

The absorption lines of this example must be due not to one galaxy, or even three: each redshift group apparently belongs to a cluster of galaxies encountered on the line of sight, each cluster with its own redshift, and each galaxy within the cluster moving with its own peculiar velocity. These clusters are, of course, so distant that they would never be seen by their emitted light.

Intervening galaxies can make their presence felt by an entirely different process: their total mass can deflect light and radio waves, and a properly placed galaxy can act like a lens, giving a magnified or a double image of the quasar behind it.

Gravitational doubles

Einstein's theory of general relativity shows that rays of light can be deviated from a straight line by the gravitational field of a massive body. The first successful test of this was in 1919, when Eddington and Dyson showed that the apparent position of a star shifted by 1.7 arc sec when it was seen close to the edge of the Sun. More obvious examples have now been found in some of the distant quasars.

The first example was found in 1979 by D. Walsh of Jodrell Bank, R.F. Carswell of Cambridge and R.J. Weymann of Tucson when using the 2.1 m optical telescope at Kitt Peak, Arizona, in attempts to find optical counterparts of several hundred radio sources in a Jodrell Bank survey. One of these radio sources (0957+56 in the catalogue) was known to be close to a pair of blue star-like objects separated by only about 6 arc sec. From measurements made at Kitt Peak on 29 March 1979 Walsh, Carswell and Weymann concluded that this visible counterpart of the radio source was a kind of illusion: the two parts are the same object seen by two different light paths. The proof is that the spectra of the two parts are identical, both showing redshift $z = 1.41$.

This double quasar is a single object seen through another galaxy. Radio maps show much more detail, including the shape of the individual images. Although the intervening galaxy is very faint, with a large redshift $z = 0.3$, these observations can tell us how massive and how large it is; furthermore, they tell us that it must be in a cluster of galaxies which add to the gravitational field.

The two images of 0957+56 arrive along different ray paths, one of which may be several light years longer than the other. Further, the quasar is variable, changing by up to 0.5 magnitude in six months. The changes will be seen earlier in one image than in the other, giving a check on the geometrical model. The two images are indeed observed to vary independently, but the actual time delay has not yet been measured.

The absorption lines of the quasars, and the gravitational lens, of which several examples are now known, demonstrate the power of quasar research in exploring the Universe. On the line of sight of 0957+56 there happen to be two narrow line absorptions at $z = 1.12$ and at $z = 1.39$, as well as the lensing galaxy at $z = 0.37$. The boundaries of accessible space have indeed been pushed far out by the discovery of quasars. It is now time to turn to the study of the Universe on an even larger scale.

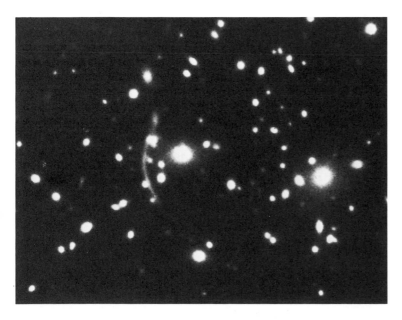

The luminous arc in the galaxy cluster Abell 370 is now believed to be an image of a galaxy produced by a gravitational lensing effect. The arc is so faint that effective study has only become possible with the introduction of electronic charge-coupled devices (CCD).

20

Cosmology

The revelation that the Universe as a whole is an entity which is accessible to rational enquiry is possibly the most important scientific discovery of this century. We now know that we can explore it both in space and in time. In space we reach beyond the solar system and the Milky Way to the galaxies and the clusters of galaxies, which we see as the basic units of the Universe. In the time domain we see change and evolution in all components, and we can now extend our comprehension to the evolution of the Universe as a whole.

Some of the basic data for this new understanding have been available since the earliest years of science. Anyone can see that the night sky is dark, and this seemed to be inconsistent with a simple model of the Universe which was infinite and unchanging. In such a model any line of sight would eventually meet a distant star, whose brightness would be similar to that of the Sun; a glance at a typical photograph of the sky shows that this cannot be happening. Arguments that there might be dark obscuring clouds beyond a certain distance destroy the simplicity of the model, and are invalid in an unchanging Universe since they would themselves be heated and shine with the same universal brightness as the stars.

Halley recognised the dilemma in 1720, but it was Jean-Phillipe Chéseaux in 1740 who suggested a solution in terms of the absorption of the light from distant stars. The problem was re-stated by Wilhelm Olbers in 1823 who also proposed a solution in terms of the absorption of light from the distant stars. The dilemma of a dark sky in an infinite universe is now generally referred to as Olbers' paradox.

Any observer who gazes at the night sky from a suitably dark site can see that beyond the stars of the Milky Way the sky is not only dark, but uniformly dark in all directions. Quantitative measurements of the most distant components of the Universe, such as counts of galaxies and measurements of background microwave radio waves, show this isotropy to a remarkable degree of accuracy. We seem to be at the centre of an evolving universe, an odd position to be in when one contemplates the fact that there are estimated to be more than a billion galaxies within the field of view of modern telescopes, and one hundred thousand million stars in each galaxy.

In the Copernican hypothesis the Earth was no longer the centre of the solar system, and the observations of William Herschel indicated that the Sun was not at the centre of the Milky Way. Today we evade the idea that we are in a central position in the Universe by interpreting one further observational fact, for which we turn to

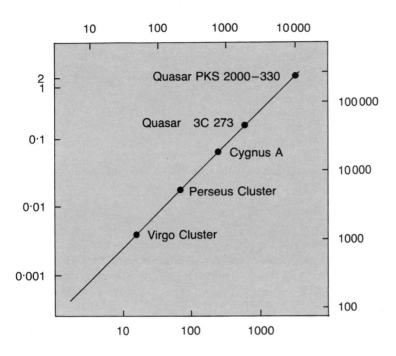

Edwin Hubble showed that the redshifts measured in the spectra of galaxies are in direct proportion to their distances. Assuming that the redshift is a result of the galaxies speeding apart, the relationship can be used to deduce the distances of the furthest objects in the Universe. The accuracy depends on how well the distances to nearer galaxies can be found by other independent methods.

Hubble's work on the redshift of the galaxies. The redshift is a direct measure of recession velocity*. Hubble's observations show that the recession velocity is proportional to distance (as measured by brightness or diameter or any other means).

Every galaxy in an expanding universe which obeys Hubble's Law sees exactly the same expansion law. It is as though space itself is expanding, and an observer in any part of space will see an equal expansion in all directions.

Whereas Chéseaux and Olbers explained the darkness of the night sky in terms of the absorption of the light from distant stars, we see now that Hubble's law accounts for the paradox. The distant galaxies were not contributing to the brightness of the sky as they would in a stationary universe, due to their velocities of recession which both shift their spectra to longer wavelengths and reduce the total flux of light in our direction. But Hubble's law gave much more: it gave the first measurement of the time scale of the Universe.

The age of the Universe

If we could turn the clock back, so that the galaxies we now see receding so fast were closer together than they now are, how far

---

* The shift $\triangle\lambda$ in wavelength λ of a spectral line from a distant galaxy may be very large, and the simple Doppler interpretation $\triangle\lambda/\lambda = z = v/c$ needs the more accurate relativistic relation

$$1 + z = \left(\frac{1 + v/c}{1 - v/c}\right)^{1/2} .$$

back must we look in time before all galaxies coincide? Note that the furthest galaxies move the fastest, so that this moment represents the birth of the whole Universe as we now see it. This time, the Hubble time, is 1.5×10^{10} years, within an accuracy of about 30%. (The uncertainty lies in the distance scale, not the velocities.)

The expansion may seem a remote and intangible concept, and the time scale may seem too long to comprehend. The time scale is, however, not unfamiliar in terrestrial terms. Let us look back through the fossil record on Earth to the earliest time when life appeared: about 3 billion years ago. At this time the scale of the Universe was about 80% of what it is now, and the density was double. The whole Universe is only about five times older than life on Earth. What was it like at its birth, when everything we now observe was condensed in a single, explosive nucleus?

Before we attempt to answer this question we should look briefly at an alternative explanation of the observed expansion, which does not require this explosive singularity in time which we refer to as the 'big bang.' The alternative is the 'steady state cosmology', proposed in 1948 by Hermann Bondi, Thomas Gold and Fred Hoyle.

The simple proposition of the steady state cosmology is that matter is created continuously throughout the Universe at a rate which exactly balances the outward flow observed in the receding galaxies. The required rate of creation is small: only a few hydrogen atoms per second in a volume the size of the Earth, so low as to be practically unobservable. The effect would be a Universe that, on average, had the same properties at all times and at all distances.

The choice between the two models, big bang and steady state, led to a famous controversy in the 1950s, when the new observational data became available from the radio astronomers.

The numbers of radio sources

When it became clear that most of the celestial sources of radio frequency radiation were at large 'cosmological' distances, a simple test could be made of the steady state model. According to this model the average population of these radio sources, whatever they are, must be the same at all times and at all places in the Universe. The total number N which could be detected with a radio telescope

Statistical counts of radio sources in relation to their strength can, in principle, be used to make deductions about the nature and history of the Universe. The counts illustrated in this graph have been adjusted so that a static Universe with an even distribution of radio sources should produce a horizontal straight line. The fact that the graph is not a straight horizontal line demonstrates that one or more of the assumptions is incorrect. An accurate interpretation is difficult to make but the observations are not regarded as consistent with a 'steady state' theory of the Universe.

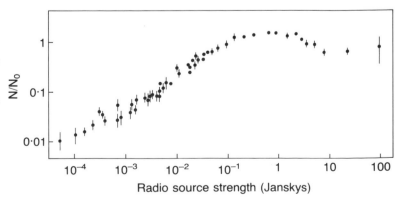

sensitive to a signal strength S would therefore follow a simple law, and it can be shown for nearby sources that N would be inversely proportional to $S^{3/2}$. For greater distances, i.e. for large red-shifts, the index 3/2 would depend on various factors, but it could never be larger than 3/2.

The radio telescope that gave the first number counts for this test was built by Martin Ryle in Cambridge, and was first used in 1953. The first catalogue of radio sources, known as the 2C catalogue, gave the very large value 3 for the index, decisively greater than 3/2 and apparently disproving the steady state theory. Unfortunately there were errors in the catalogue, which were pointed out by Australian radio astronomers and seized on by the proponents of the steady state theory. The next catalogue, 3C, was free of the faults of 2C, and gave an index of 1.8, above 3/2 but not so decisively. Subsequent catalogues, however, have given radio source counts that cannot possibly be reconciled with the steady state theory. As S decreases, the number of sources at first increases, then falls by a large factor for the weakest, most distant radio sources.

This early work on radio source counts is now supported by more discriminating studies of different categories of objects, such as the quasars. These are evidently very rare objects, and as such the

This radio telescope was built at Cambridge in 1953 and was used for the survey on which the Second Cambridge (2C) Catalogue of radio sources was based.

ones we detect are likely to be at very great distances. There are several known with redshifts z greater than 3.5, but a notable lack at redshifts greater than 4. This population deficiency is not merely an illusion caused by the difficulty of detecting quasars at such a distance: it is a real deficiency, indicating that the epoch of quasar activity in galaxies starts at a definable and very early stage in the life of the Universe.

The cosmic radio background

Looking back towards the big bang we expect to see a Universe which is progressively hotter as the average density increases. The measured recession velocities of the galaxies, used for establishing Hubble's law, are now sufficiently accurate for us to say how small or how hot the Universe was at the time of the big bang. If we had any indication of physical conditions soon after the big bang, we might be able to be more precise about this important question. Surprisingly we do know something about a very early stage, at an epoch which we believe to be less than a million years after the beginning of the expansion.

The temperature at that time was about a few thousand degrees. At that epoch the temperature had fallen to a level at which hydrogen was no longer ionised, and matter, instead of radiation, became the dominant feature of the expanding universe. The present day temperature depends only on the scale of the expansion. Today the temperature of this cosmic radiation is 2.8 K, which means that it is detectable at microwave radio wavelengths.

The original measurement of this relict radiation from the big bang was the fortuitous outcome of some very careful radio engineering work by Arno Penzias and Bob Wilson at the Bell Telephone Laboratories in the USA. They were testing a very sensitive radio system for 7 cm wavelength, intended for radio communication via the *Echo* balloon satellite. Working with a horn antenna, they found that their receiver was picking up a weak radio signal from whatever direction the horn was pointing in. This first detection was followed by others, over the wavelength range 50 cm to below 1 mm. The spectrum over this range was exactly the blackbody curve expected for a temperature of 2.8 K.

This discovery is convincing evidence for the big bang model, taking the observations back to an era a few billions of years before the formation of the galaxies. Can we look even further back to a time of even higher temperatures, closer to the big bang itself?

The formation of the elements

Although the spectrum of starlight may be dominated by the spectral lines of elements heavier than hydrogen and helium, these elements are in fact comparatively rare in the Galaxy taken as a whole, amounting to no more than 1% of the total mass. Apart from this rare but important population, it is a good approximation to say that about one atom in ten is helium, the rest being hydrogen. By mass, the accepted values are 23% helium, 77% hydrogen. The same ratio seems to apply in other galaxies, and we may ask

whether this composition can be accounted for by the physical conditions at the time of, or immediately after, the big bang.

The first analysis of nuclear reactions in the hot conditions of the big bang was by George Gamow. He found that the natural mixture to be expected contained only hydrogen, deuterium and helium. The present mixture, containing heavy elements, could not be primordial, and possibly told us nothing about the early universe. Fortunately the origin of the heavy elements was soon discovered: they were created at a comparatively recent era, in stars and in supernova explosions.

The synthesis of heavy elements in a normal star is a natural consequence of the nuclear reactions which are the basic source of the star's energy. For most of the lifetime of a normal star hydrogen is converted to helium; at later stages the helium combines to form carbon, and then oxygen, neon, magnesium and silicon in succession, each species containing one more helium nucleus. Elements heavier than iron cannot be made in this way: these can only be created in a supernova explosion, as described in Chapter 16.

The theoretical work on nucleosynthesis gave a complete explanation for the abundance of the heavy elements, and the field was open again for a cosmological explanation of the relative abundance of hydrogen and helium. Gamow's work showed the way in principle, and it has more recently been found that the hot big bang gives exactly the right proportion of hydrogen and helium. The key

The horn antenna at Bell Telephone Laboratories used in 1963 by Arno Penzias and Robert Wilson in the discovery of the cosmic background radiation.

moment at which this proportion was determined was 100 s after the beginning of the expansion. At this time the temperature was 10^9 K, and the density had already fallen to one-thousandth of the density of water.

The observations of helium abundance in our present day universe allow us, therefore, to determine the physical conditions only 100 s after the birth of the Universe. Does it make any sense to look back any further in time?

The first second of time

The calculation of the helium abundance was possible because ordinary laboratory physics could be applied to the conditions predicted to exist when the Universe was 100 s old. Before this time, say at time 1 s, the temperature must be so high that only the particle physics of high energy accelerator machines can have any relevance. Again we need to ask if the simple model of expansion from a point is reasonable: perhaps the random element of velocities seen today indicates a real limit to the density from which the expansion started. There is, however, no escape along these lines, as was shown by Stephen Hawking in 1966. In general relativistic theory the force of gravity is so strong that the singularity, or point-like nature of the infant Universe, cannot be modified by random velocities. We are therefore in the realm of particle interactions at the highest imaginable energies.

Remarkably, this region of cosmological theory is closely related to modern particle physics. The forces between elementary particles, apart from gravity, are of three types: electromagnetic, strong nuclear, and weak nuclear, all of which obey different laws of variation with distance. In 1967 Weinberg and Salam developed a unified theory which integrated the electromagnetic force and the weak nuclear force. The unification into the electroweak force was predicted to require the existence of hitherto unknown particles. Subsequently in the high energy accelerator at CERN these particles – known as the W and Z particles – were discovered. In the condition of the early Universe it is believed that only one single force existed, but the theory required to unify the electroweak and the strong nuclear force has not yet been evolved. Finally, the formidable problem of a quantum theory of gravity has to be solved before we can be confident that we understand how the four fundamental forces of nature were unified in the initial moments of the Universe.

The future of the Universe

Our description of the recession of the galaxies, which we regard as indicative of the expansion of the whole Universe, lacks one essential component. We know the velocities, but we cannot say if they are constant, or increasing, or decreasing. Gravitation is the predominant force at all large distance scales, so we would expect to see the expansion slowing down under the mutual gravitational force of the galaxies themselves. This force, averaged over a large scale, depends on the mean density ρ of the Universe; above a certain value ρ_0 the expansion will eventually be halted and reversed, leading to a collapse – the 'big crunch'; for a density below ρ_0 the expansion will never be halted.

The value of ρ_0 is simply calculated: it depends on the slope of the Hubble graph and the value of the gravitational constant. The best estimate is $\rho_0 = 10^{-29}$ g cm^{-3}. It would be a simple and appealing situation if the mean density of the Universe were found to be exactly equal to ρ_0, so that the expansion could be expected to slow eventually to zero velocity but not to reverse.

The total mass of matter in our Galaxy, and in other galaxies and in clusters of galaxies, can be found from their dynamics, assuming that their kinetic energies are balanced by gravitational forces. Surprisingly, the resulting value of ρ differs from ρ_0 only by a factor of ten. The matter we can detect is insufficient to 'close the Universe'; perhaps there is other material which we cannot detect, a 'missing mass' in some unsuspected form. Present indications are against this hypothesis, and we are left with an irreversible, inexorable expansion of the Universe, with a time scale that has a beginning but no end.

Whence the structure?

Most of this chapter has dealt with the Universe as though it had no detailed structure, and could be described simply by general characteristics such as age, distance scale, density, and temperature. But on the small scale of the reader and this book there is very detailed structure, and we have been concerned throughout the book with structure on the scales of the planets, the solar system, stars, the galaxies, and clusters of galaxies. Presumably we should approach the question through the theory of turbulence, in which structure appears first at the largest scales, and breaks up later into smaller and smaller scales. But we do not know how the large scale structure appeared, or even at what stage in the expansion it appeared. Obviously it had appeared before the epoch of formation of the quasars, before the time corresponding to $z = 4$, i.e. when the age of the Universe was about one billion years; equally obviously it had not appeared when the microwave background radiation became detached from the material universe, at a time of about one million years after the beginning of the expansion.

The only observations which may be relevant to this problem are those which are searching for some slight asymmetry in the microwave background radiation, on a scale which might indicate an appropriate incipient structure in the material Universe at the time when the background radiation began to expand independently of the matter. Several experiments are underway to search for this; so far the result is that there are no irregularities in intensity above about one part in 10^4 on any relevant angular scale, and none above two parts in 10^5 on the angular scale of a few degrees where structure is most expected.

Cosmology has indeed made good progress from the original observation that the sky is dark. We now know that we live in a rapidly changing Universe, that we can at least aspire to understand the formation of the galaxies, and that we have a glimpse of the youthful Universe as it must have been soon after the big bang.

21

The origin of life

The only available evidence about living organisms exists on Earth. Do we know when the simplest living organisms appeared on Earth? Microstructures have been found in ancient rock formations which have been interpreted as the fossils of single-celled organisms. The rocks are 3.2 billion years old. If the interpretation is correct, that these structures are the fossils of living cells, then life must have emerged in Earth 3.5–4 billion years ago, that is 0.5–1 billion years after the Earth itself formed from the solar nebula. Since the publication in 1953 of the classic paper by Watson and Crick, the significance of the DNA molecules (deoxyribonucleic acid) in determining evolution from the simplest living organism has been widely accepted by the scientific community. The remarkable ability of this DNA molecule to copy itself is the essential factor determining inheritance. Once the DNA molecule existed the cell could divide with each new cell possessing a copy of the DNA molecule and thus resembling the parent. Evolution then occurs through random alterations of the DNA strand – and only those changes which give rise to beneficial interactions of the organism with the physical environment enable the organism to survive. This is believed to be the underlying scientific basis for the process of natural selection described by Darwin in 1859.

The transition from non-living to living matter

If the emergence of ourselves from the simplest living organisms over a period of 3.5–4 billion years occurred through evolutionary processes of this type, the critical question remains concerning the origin of the complex molecules and their transition to the simplest living organisms in the first billion years of the history of the earth.

Three different attitudes are prevalent over this vital question. First, and particularly in the USA, there are recurrent strong bodies of creationists who maintain that this is not a scientific question but that life originated on Earth through a divine act. Second, there is the view most recently pressed by Hoyle that the simplest organisms are widespread throughout the Universe and have been transported to Earth by some natural means. In its ancient form this was the doctrine of panspermia. At the beginning of this century the theory of lithopanspermia had many adherents. This was the theory that the germs of life or complex living organisms were brought to earth by meteorites or cosmic dust. In a variant of this theory the Swedish scientist Arrhenius proposed that a physical carrier was unnecessary and that the organisms were transported by the pressure of light. These theories had to be rejected when it was demonstrated that living organisms could not survive the ultraviolet radiation to which they would be subject during their journey through space. The contemporary revision of this type of theory by Hoyle overcomes this problem. He bases his arguments on the assertion that the organisms are protected from any damaging radiation by a coating of carbon and claims that absorption spectra in the ultraviolet obtained in space vehicle observations demonstrate that particles of the appropriate size exist in space. In favour of these arguments Hoyle points to the large number of molecules that are known to exist in the interstellar clouds. The presence of over 50 molecules, including water, has been revealed by the observation of the spectral lines in the short wave radio region of the spectrum. In common with a number of scientists Hoyle believes that in the interstellar clouds the necessary molecular material exists for the formation of the amino acids and nucleotides that are essential for the evolution of living organisms. The problem of transferring these organisms to Earth from an interstellar cloud remains. Hoyle's solution is a reversion to a form of lithopanspermia – that the major journey through space takes place in the icy conglomerate of a comet and the final transfer to Earth is by meteors, micrometeorites or meteorites that are associated with the comet.

On the whole, conventional scientists have been very critical of this modern panspermia theory of the origin of life from space. On the other hand the theory that living organisms exist in space is consistent with the widespread belief that forms of life exist elsewhere in the Universe. Indeed in 1985 details were published of experiments carried out by scientists in Leiden which appear to show that the original arguments that destroyed the theory of panspermia may not be correct. In the Leiden experiments bacterial spores survived in a laboratory environment similar to that to be expected in space – in a vacuum chamber in which the temperature had been reduced to near absolute zero and exposed to intense ultraviolet radiation. The scientists estimated that within the protective environment of an interstellar cloud and when the spores became coated with the protective skin of the type proposed by Hoyle the bacterial spores might survive for several hundred million years. Although the theory that the simplest living cells originated in space and were transported to Earth by some means may be regarded with reserve, it does not appear that any conclusive arguments have yet been produced against this idea of the extraterrestrial origin of the life that has emerged on Earth.

The third attitude – that life on Earth emerged from non-living matter – has been the most favoured one in recent years. In 1870 T.H. Huxley promulgated his opinion that life originated in lifeless matter, but the revolution in outlook which gave this theory a scientific basis began in the 1920s and is associated with the work of A.I. Oparin in the Soviet Union and J.B.S. Haldane in Great Britain. The basis of the theory is that complex organic molecules developed from the gases in the primeval atmosphere of the Earth with solar radiation or electrical discharges in the atmosphere providing the energy sources. A striking experiment in favour of this view was carried out in California in 1952–3. The gases believed to be prevalent in the primordial atmosphere of the Earth are hydro-gen, ammonia, methane and water vapour. At the suggestion of Harold Urey his student, S. Miller, mixed together these gases and circulated the mixture through an electrical discharge. After a week of this treatment the water was found to contain several types of amino acids. In further experiments of this type other complex molecules have been produced including nucleotides. Also other forms of energy including ultraviolet radiation were found to be effective in the synthesis.

If these processes occurred in the primordial atmosphere of the Earth the complex organic molecules would have been dissolved in the oceans, and during the course of the first millions of years of the Earth's history a dilute concentrate of amino acids and nucleotides would have accumulated in the primeval seas. If the interpretation of the fossil records of the ancient rocks is correct then the first living cells emerged on Earth less than one billion years after these complex molecules accumulated in the seas. The question then con-cerns the transition by which these organic molecules formed into the very large complex molecules, to produce about 20 different kinds of amino acids and the five nucleotides now known to play a critical role in living organisms – that is in the formation of proteins, enzymes, and the nucleic acids including DNA.

Although this summarises the essential process by which life is believed to have emerged on Earth it must be said that our know-ledge of the transition from the non-living complex macro-molecules in the early oceans to the primitive single-celled organisms is speculative and fragmentary. As described above there is scientific evidence that complex molecules including the amino acids could form from the gases present in the primordial atmo-sphere, and after the first single-celled organisms evolved there is a combined geological and biological explanation of the evolutionary processes that led to our existence today. However, any discussion of the transition process from the non-living to the living era is beset with immense complications and difficulties. For example, given that the amino acids and nucleotides existed in the primeval oceans it has been estimated that some 10^{130} possible alternative sequences and combinations are possible. For life to emerge the correct sequence must have occurred within a time span of 0.5–1 billion years. How did the critical and unique selection of the combination of molecules occur out of this gigantic range of possibilities? The probability of a chance occurrence of the correct sequence is vanishingly small, and in recent years a great deal of research has been concentrated on the conditions that may have op-erated to raise the probability to a finite level.

Many scientists most deeply immersed in this problem fully accept that the necessary complex molecules could have formed in the early oceans, and that man evolved from the simplest living cell over a period of 3.5 billion years. However, they consider that the probability of the transition from the non-living molecules to the living cells within the boundary conditions of time and space on Earth is so vanishingly small that the only correct scientific attitude to assume is that the transition did not occur on Earth. It is this understandable scientific attitude that has led biologists such as Crick to arrive at conclusions, similar to those popularised by

Hoyle, that the transition from non-living to living material could have occurred only in the infinitely greater boundaries of time and space of the Universe and that living material was transferred to Earth by some process as already discussed.

Does life exist elsewhere in the Universe?

The possible existence of a multiplicity of worlds in the Universe has for centuries been the subject of philosophical speculation. When the work of Shapley and Hubble in the 1920s revealed the extent of the Milky Way and the multiplicity of similar galaxies a scientific formulation was established for such speculations. At the time of these discoveries Harlow Shapley, who was then the director of Harvard Observatory, concluded that there must be some one billion stars in the Universe with characteristics similar to those of the Sun which would have a planet with the appropriate physical and other conditions like Earth favourable to the emergence of life. However, in that era, the encounter theory of the formation of the solar system was dominant. This was the theory that the solar system was formed when a star passed close enough to the Sun to pull out from it by gravitational attraction the mass of gas and dust that remained as a nebula around the Sun and eventually condensed into the planetary system. James Jeans, one of the eminent and influential astronomers of that period and a proponent of the encounter theory, argued that, whilst not questioning Shapley's estimates of the number of stars in the Universe similar to the Sun, nevertheless the solar system and life on Earth must be unique. Jeans calculated that the probability that two stars could approach one another in such an encounter would be vanishingly small; in other words that the formation of the solar system was the result of a unique event in the Universe. He also maintained that on the encounter theory the system of planets would have formed in a molten condition from the high temperature material pulled out from the Sun by the passing star, and therefore life on Earth could have arisen only at a later stage in its evolution when it had cooled sufficiently for life forms to exist.

In the upsurge of astronomical interest and discovery after World War II the scientific basis of these conclusions that appeared to endow the Earth and life with a high probability of uniqueness in the Universe was undermined. In the modern era there has been an almost complete reversal of opinion for two reasons. First, the encounter type theories are regarded as untenable, and the processes of star formation from interstellar clouds favour the view that the majority of stars will evolve with a surrounding nebula of gas and dust and that, as in the case of the Sun, planets can evolve from this nebula. This means that planetary systems and stars would be expected to be a common feature, and in this respect the solar system is not unique. Second, the molecular constituents from which the amino acids and nucleotides could form have been shown to exist in the interstellar clouds from which the stars evolve.

Because of these developments a much firmer scientific basis has existed in recent years on which a discussion of the possible existence of extraterrestrial life can be based. Even on the assumption that planetary systems around stars are a common feature, an estimate must first be made of the number of stars that have sufficient long term stability like the Sun. The conclusion is that stars not too dissimilar from the Sun are the most likely candidates in the spectral range F to K (the Sun is spectral class G) with temperature in the range 4500 to 7500 K. These include about one-quarter of all the stars in the Milky Way, and when allowances are made for binary and multiple star systems this leads to the estimate that about 20 billion stars in the Milky Way could, in principle, possess a planet with Earth-like characteristics. As far as the observable Universe is concerned, even if only spiral galaxies similar to the Milky Way are considered, the conclusion appears to be that many tens of billions of stars could have planetary systems with the condition of long term stability necessary for the emergence of life.

The weakness of this conclusion is that no planetary system has yet been observed around even the stars nearest to the Sun. Of course, in view of the limited resolving power of telescopes coupled with the difference in brightness between star and planet this is not unexpected. However, there is an accumulation of circumstantial evidence from the observed irregularities in the motion of some of the nearer stars, that these stars have planetary systems. Future observations from space could well give decisive evidence on this crucial issue.

On the basis of this circumstantial evidence and with confidence that the process of star formation from the interstellar clouds is understood, at least in principle, contemporary opinion is that there must be billions of stars in the Universe with planetary systems under stable conditions for several billions of years. The next critical question to be considered is how many of this multiplicity of planets are likely to have developed habitable atmospheres? We have already described in Chapter 9 the narrowness of the zone and the extreme delicacy of the conditions in the solar system that led to the differences between Earth and Venus and Mars. Only on Earth has a habitable atmosphere emerged through a series of delicate balances over billions of years, and in these balancies the emergence of primeval organisms and plant life has been a critical factor. Although we have evidence for the critical nature of the conditions we do not know what the limiting factors are even for our own solar system.

At present this seems to be one of the weaker links in the scientific discussion. If we do not know these limits for our own system how can we possibly assess the probability that a planet around another star will have developed a habitable atmosphere? As yet there is no agreed scientific response to this important query. One view is that the distribution of planetary distances around stars of differing luminosity may be in a similar ratio to that found for the Sun. In this case it was argued that there would be the same probability of finding one of the planets within the appropriate distance and with the other conditions necessary for the evolution of a habitable atmosphere. Evidently further observational evidence is required before the scientific basis of these discussions can be strengthened. In this respect the knowledge of the conditions on another planetary system would be of great significance.

The search for extraterrestrial intelligence – Project SETI

Encouraged by the considerations outlined here considerable sections of the astronomical community accept as axiomatic that life exists elsewhere in the Universe. Equations expressing the probability of the existence of extraterrestrial life have been developed. Many of the terms in these equations, such as the probability that a planet will develop a habitable atmosphere, are assessed at a low figure. Even so, the tens of billions of stars with the potential for possessing planets raise the final assessed probability to a finite value. Considerable sums of money have been committed, particularly in the USA, to the development of equipment with which the search for signals from extraterrestrial life are being pursued. The importance and scientific validity of these developments were underlined by the formation in the 1970s of a special commission of the International Astronomical Union.

The scientific case for a search was made in a letter received by one of the authors (BL) on 29 June 1959 from Giuseppe Cocconi of Cornell University, who was at that time on leave at CERN in Geneva. In this letter he urged that the large radio telescope at Jodrell Bank which had recently become fully operational should be used in the search for intelligent extraterrestrial signals. In collaboration with his colleague Philip Morrison the arguments were published a few months later in *Nature*. After outlining the points made in this chapter Cocconi and Morrison expressed the view that, if life had developed on extraterrestrial planetary systems, then among such systems there might be some societies which had maintained themselves for a very long time compared with the time scale of human history. In such cases the societies might have scientific interests and techniques much in advance of our own and that they would have recognized our own planetary system as a probable place for life to develop and would long ago have transmitted signals in the hope of eliciting a response. Cocconi and Morrison made the important suggestion that they would have carried out these transmissions on an objective standard of frequency throughout the whole Universe – that of the spectral radio emission line from neutral hydrogen on a wavelength of 21 cm. On the assumption that the transmitting facilities of the extraterrestrial society were at least as advanced as those on Earth Cocconi and Morrison had calculated that with the techniques available in 1959 suitable receivers used with radio telescopes of the size of the instrument at Jodrell Bank should be able to detect signals from communities existing at the distance of the nearer stars.

Subsequently many papers have been published on these issues and on the means of establishing communication. Preliminary but negative searches were made by radio telescopes at the National Radio Astronomy Observatory in the USA and by scientists in the USSR. Increasing pressure was exerted for the construction of special radio systems in order to engage in a comprehensive search and this culminated when a multi-million dollar project – SETI – was initiated in the USA, using the most advanced computerised techniques.

SETI, the search for extraterrestrial intelligence, is a formidably difficult project. If there is a detectable radio signal arriving on Earth, it will be weak, we will have no idea in which direction we should point the radio telescope to receive it, and we will have little or no idea what radio frequency to tune our receivers to. The characteristics of the signal will be very simple: most probably it will be a pure unmodulated oscillation. It will be so weak that several minutes of observing will be required for each frequency and for each direction of pointing the telescope. The only hope of completing a search in a reasonable number of years is to use a multi-channel receiver, and carry out the search in an automatic, continuous sequence of observations.

A severe practical difficulty is to recognise and reject all signals of terrestrial origin. In our civilised society we are encompased by radio signals of all kinds, including many transmissions accidentally spreading into the quiet frequency bands reserved for radio astronomy. An early test of SETI was therefore made in 1983 at Jodrell Bank. A small radio telescope, 12.8 m in diameter, was assigned to the project for two months. It was connected to a prototype SETI receiver, with 64 000 channels, working in the decimetre wavelength radio band. The task was to list all the interfering sources which could be located near Jodrell Bank and to assess the chances of seeing a weak extraterrestrial signal against that formidable background. SETI is now extending its receiving techniques, and the next tests will use a million-channel receiver.

We may speculate on the next moves if a signal is identified as extraterrestrial. First, it must be confirmed using a completely different set of apparatus on a different radio telescope, to eliminate the possibilities of error and even of deliberate deception by a local radio enthusiast. Next, how should we respond? The chances of a real communication are very low, as the transmission time will be tens or hundreds of years. Neither the sender nor the receiver will be expecting an acknowledgment; if we do reply, we need a very powerful transmitter that must be devoted to the task for many years, with almost no chance of assessing its success.

There are those who even consider it dangerous to acknowledge reception of any extraterrestrial signal, since the civilisation that sent it might be malevolent. There is, however, no hope of hiding our presence, as we have over the last half century devoted very large resources to building powerful transmitters of radio, television and radar signals. We are a new kind of radio nova, possibly already detectable by a SETI project on another planet. Apart from these random transmissions a special coded message was transmitted from the Arecibo radio telescope in 1974 directed at a cluster of stars which the message would reach in the year 30 000 AD. The code was the modern analogue of Morse code using the digits 0 and 1 instead of dots and dashes. If successfully translated by an extraterrestrial community the picture would have revealed the Arecibo telescope, a diagram of the Earth, Sun and planets, a human form, the double helix of the DNA molecule and the appropriate chemical formula. Four American space probes (Voyager 1 and 2, Pioneer 10 and 11) having completed their missions in the solar system are now travelling in interstellar space. In 80 000 years time they will have covered a distance equal to that of the nearest star from the

Sun. The Pioneer probes contain messages in diagramatic form on plaques, the Voyagers carry the messages on a video disc. Should these probes ever be recovered by an intelligent extraterrestrial community these messages will reveal some fundamental features of 20th century civilisation on Earth.

The special position occupied by man

Any measure of success in the search for extraterrestrial intelligence would, of course, have a revolutionary impact on man's attitude to his relation to the Universe. A vast area of scientific uncertainty would be immediately swept aside. Nevertheless, although the contemporary SETI investigations use the most modern techniques the penetration into space now, and for the forseeable future, will encompass relatively few of the billions of stars in the Universe that are considered to be possible candidates possessing habitable planetary systems. The probability is, therefore, that the search will yield negative results and will leave open the whole question of the existence of extraterrestrial life. The various astronomical and biological considerations have wide areas of uncertainty, and the individual assertions and beliefs about the uniqueness of life as we understand it on Earth may remain a matter of personal preference. On the one hand the observable Universe seems to contain all the necessary features for the widespread distribution of life. Many billions of stars around which planetary systems with long term stability probably exist, and the interstellar clouds contain the necessary molecular constituents which could form the basic material from which replicatory cells develop. On the other hand the Earth appears as a most delicately balanced system even amongst the nearest planets of physical similarity in the solar system. In particular the circumstances through which the Earth lost the primordial atmosphere to be replaced, probably through volcanism, with the habitable atmosphere in which man has evolved, remains entirely unknown. We have no scientific evidence to indicate whether catastrophic events of that type would inevitably occur on other embryonic planetary systems.

Until many of these issues are settled either by observation of other planetary systems, or by direct contact with an extraterrestrial intelligence through the present SETI equipment or its successors, a rational attitude is that man occupies a special position in the Universe. Although we understand the biological sequences through which man has evolved from the simplest living organisms we do not know whether consciousness is an automatic consequence of such evolution. This may be the ultimate assumption, since through our consciousness we contemplate and comprehend the Universe.

Glossary

albedo The reflecting power of objects such as the Moon and planets defined as the fraction of the light energy falling on a surface that is reflected.

altazimuth mount A type of telescope mounting that allows the tube to be moved independently in altitude (height), by swinging around a horizontal axis, and in azimuth (position round the horizon), by swinging around a vertical axis.

Andromeda galaxy A large, nearby spiral galaxy. The only one that can be seen with the naked eye.

angular momentum The momentum of rotation, defined as the product of angular velocity and moment of inertia.

angular velocity The rate at which angular distance is covered or rotation takes place. The proper motion of a star across the celestial sphere might be measured in arc seconds per century and the rotation of a neutron star in revolutions per second. Both are examples of angular velocity.

arc minute A unit used for measuring the size of small angles equal to one sixtieth of a degree.

arc second A unit used for measuring the size of very small angles, equal to one sixtieth of an arc minute or 1/3600 of a degree.

astrometry The branch of astronomy that deals with making measurements of the positions of objects in the sky.

astronomical unit (AU) The mean distance of the Earth from the Sun, approximately equal to 150 million kilometres or 93 million miles.

aurora Luminous curtains or streamers of light seen in the night sky at high northerly or southerly latitudes, caused by electrically charged particles from the Sun streaming into the Earth's atmosphere, guided by its magnetic field.

Cassegrain focus An optical arrangement, commonly used in reflecting telescopes, which makes use of a hole in the centre of the primary mirror. A convex secondary mirror near the top of the tube reflects the light from the primary back towards the hole. The light is brought to a focus just behind the primary mirror.

Coriolis force A force that arises in problems involving rotation, when a moving object is changing its distance from the centre of rotation, such as when a projectile is launched from the rotating Earth. It is similar in concept to the centrifugal force felt when a vehicle goes round a corner.

coronagraph An instrument for producing an artificial eclipse in a solar telescope so that the corona can be observed.

couple The effect of forces acting on an object that tend to make it rotate rather than move in a line.

diffraction grating An optical device that produces a spectrum when light passes through it or is reflected from it. It is made by engraving very close, straight parallel lines on, for example, glass or metal.

dipole antenna A simple rod-like aerial for collecting radio waves.

Doppler effect The change in the observed frequency of light or sound waves when the source of the waves and the observer are moving either towards or away from each other.

electromagnetic radiation A form of energy, with wave-like properties, consisting of oscillations in linked electric and magnetic fields. The character of the radiation depends on its wavelength. From long to short wavelengths, the different names given to electromagnetic radiation are radio waves, infrared radiation, visible light, ultraviolet light, X-rays and gamma-rays. Under some physical circumstances, the radiation has particle-like properties, as if the energy exists only in discrete 'packets', which are termed photons.

electron An elementary particle, found in normal atoms, with a negative electric charge of 1.602×10^{-19} coulomb and a mass of 9.11×10^{-23} gram.

electron volt (eV) A small unit of energy, frequently used in atomic physics, defined as the energy acquired by an electron as it falls through a potential difference of 1 volt. It is equal to 1.602×10^{-19} **joules.**

epicycle A small circle, the centre of which travels along a larger circle called the deferent. It is a device used in trying to give a description of the movements of the planets in terms of circular motion only.

equatorial mount A method of mounting a telescope so that it can move independently in the directions of the celestial coordinates, right ascension and declination. Movement in right ascension is about an axis set parallel to the Earth's rotation axis. This type of mounting makes it easy to follow the movement of the stars across the sky either manually or by driving the telescope with a motor.

flux A general term for the amount of energy or number of particles passing through one unit of area in one unit of time. It can also be a measure of the strength of a force field (e.g. magnetic) through an area.

gravitational constant A factor, G, that links the strength of the force of attraction between two bodies (F) with their masses (m and M) and separation (x), such that $F = GmM/x^2$. Its value is $6.67 \times 10^{-11}\,\mathrm{N\,m^2\,kg^{-2}}$

gauss (G) A unit in which the strength of magnetic fields is measured. One gauss $= 10^{-4}$ tesla (T).

hertz (Hz) The unit in which frequency is measured. One hertz is one wave-cycle per second.

interferometer A telescope that depends for its operation on the combination of electromagnetic waves that originate from the same source but have travelled through slightly different paths before being collected. The method is used particularly in radio astronomy, where pairs of radio telescopes are linked in order to achieve high angular resolution.

isotropy An isotropic medium has the same properties in all directions.

joule (J) A unit of measurement for energy. One joule $= 10^7$ erg.

kelvin (K) The unit in which temperatures are measured on the scale in which zero is absolute zero. The kelvin is also the unit for measuring temperature differences and one kelvin is equal in size to one degree on the Celsius scale.

light year A unit of distance, frequently used in popular astronomy, equal to the distance light travels in a vacuum in one year, or 9.46×10^{12} kilometres.

limb The outermost edge of the visible disc of the Sun, the Moon or a planet.

Lyman alpha One of the principal lines in the ultraviolet spectrum of hydrogen, with a laboratory wavelength of 121.6 nanometres.

magnetic storm A disturbance in the Earth's normal magnetic field caused by an influx of electrically charged particles from the Sun.

magnitude A measure of the brightness of an astronomical object. The magnitude scale is based on ratios of energy output, so a first magnitude star is a hundred times brighter than a sixth magnitude one. Different kinds of magnitude (e.g. visual or photographic) can be defined according to the range of wavelength covered.

mass spectrometer An apparatus that can separate out different atoms and isotopes according to their atomic weight and so be used to find out what elements and which of their isotopes are present in a sample of material.

meridian A circle on the celestial sphere that passes through the north and south poles and the zenith. It is essentially the north-south line across the sky.

micron An abbreviation sometimes used for the unit of measurement now correctly called a micrometre, equal to one millionth of a metre.

molecular band A feature in the spectrum produced by a molecular (as opposed to atomic) substance, consisting of large numbers of closely-spaced lines.

moment of inertia The property of a rotating body that is analogous to the mass of a body travelling linearly. The moment of inertia depends on the axis about which the body is rotating and reflects the distribution of mass around that axis.

nebula The Latin word for 'cloud' used loosely in astronomy to describe any misty or extended source of light in contrast to the point-like image of a star. In a stricter sense, it is used to describe only features of gas and dust rather than clusters or galaxies composed of stars.

neutrino A stable elementary particle with no electric charge and believed to have almost no mass. Neutrinos hardly interact at all with other matter and so are difficult to detect.

neutron star A star composed almost entirely of neutrons, which form as the core of a dying star collapses inwards. Neutron stars have a mass of one or two solar masses contained in a diameter of only a few kilometres. Pulsars have been identified as being neutron stars.

nucleogenesis, nucleosynthesis The creation of new chemical elements in nuclear reactions that occur naturally, for example inside stars and in supernova explosions.

nutation A small 'wobble' in the direction of the Earth's rotation axis caused by the gravitational pull of the Moon on the Earth. The wobble is superimposed on the steady precession of the Earth's axis, which sweeps out a cone in space over a period of 26 000 years.

optical light fibre A flexible glass fibre that is able to act as a 'light guide', and thus direct light along a curved path.
A bundle of such fibres can be used to transmit an image.

Orion Nebula A glowing cloud of interstellar hydrogen gas visible to the naked eye as a misty patch in the constellation Orion.

pascal (Pa) A unit in which pressure is measured, equivalent to a force of one Newton acting over an area of one square metre. One pascal = 10^{-5} bar. (Normal atmospheric pressure is approximately 1 bar.)

plasma A hot gas consisting of electrically charged particles, rather than neutral atoms. The particles are negatively charged electrons and positively charged ions, i.e. atoms from which electrons have been stripped because of the high temperature.

Pleiades An open cluster of young stars in the constellation Taurus, also called 'the Seven Sisters', which is readily visible to the naked eye.

potential energy The energy an object has by virtue of its position in a field of force. For example, an electrically charged object has potential energy in an electric field and a mass has gravitational potential energy when it is in a gravitational field. The potential energy is released when the object moves through the field.

precession The slow change in the direction of the Earth's rotation axis, which sweeps out a cone in space in a period of 26 000 years. Precession is caused by the action of the Sun's gravity on the non-spherical Earth.

proper motion The slow change in the relative positions of the stars in the sky due to their true motion through space, perceptible only over periods of decades or centuries.

proto-planet The initial condensation of material that will eventually become a planet.

pulsar A cosmic radio source from which rapid regular pulses of radiation are detected. The time between pulses may be anything between a few seconds and milliseconds. Pulses have been identified as neutron stars resulting from supernova explosions. The pulses arise from a beam of radiation emitted by the rapidly rotating star.

quasar A distant, compact, object that looks starlike on a photograph but has a redshift characteristic of an extremely remote object. The word is a contraction of 'quasi-stellar object'. Quasars were discovered in 1963 as the optical counterparts of powerful extragalactic radio sources. Their high redshifts correspond to velocities approaching the speed of light. The implied distances run into billions of light years, making them the most distant and the most luminous objects in the universe.

radian A unit in which angles are measured. There are 2π radians in a complete circle of 360°, i.e. one radian is the angle for which radius and arc length are equal.

radical A group of atoms, bound together chemically, which can pass through a series of chemical reactions unchanged but is normally unable to exist separately. An example is OH, which is part of water and alcohol molecules, but which is found on its own in interstellar clouds.

radio galaxy A galaxy that is an intense source of radio waves. The radiation comes from two sources, one each side of the galaxy and is generated by electrons travelling through a magnetic field at a speed nearly that of light.

redshift The displacement of features in the spectra of astronomical objects, particularly galaxies and quasars, towards longer wavelengths. The effect is usually interpreted as the observable result of the Doppler effect, because the objects are receding rapidly.

relativity, special and general theories of Theories developed by Einstein and published by him in 1905 and 1916. Important features of special relativity are that the speed of light is the same for all observers, independent of any velocity the source may have, and that space and time must be regarded as components of a four-dimensional spacetime. In the general theory, Einstein sought to unify the idea of spacetime with gravitation and predicted that the presence of matter can 'bend' spacetime.

seismometer An instrument for detecting and measuring the shock waves due to earthquakes and similar phenomena.

Shuttle spacecraft (Space Shuttle) A reusable space launch vehicle, developed by the United States, which takes off like a rocket but lands like an aircraft on a runway.

spectral type A classification assigned to a star according to the appearance of its spectrum. Spectral types are correlated with the colours, temperatures and masses of stars.

stratosphere The rarified region of the Earth's atmosphere above the denser layer near the surface.

supernova A stellar explosion that occurs at the end of the life of a massive star.

tonne A unit of mass equal to 1000 kilograms. One tonne = 0.9842 ton.

torque A turning force.

transit circle A telescope designed to observe and time the passage of stars across the meridian.

troposphere The densest layer of the Earth's atmosphere close to the surface and extending upwards for between 8 and 16 kilometres.

Useful addresses

United Kingdom

British Astronomical Association
Burlington House
Piccadilly
London W1A 9AG

Junior Astronomical Society
c/o 36 Fairway
Keyworth
Nottingham NG12 5DU

United States

American Association of Variable Star Observers
25 Birch Street
Cambridge
Massachusetts 02138

Association of Lunar and Planetary Observers
PO Box 16131
San Francisco
California 94116

Astronomical Society of the Pacific
1290 24th Avenue
San Francisco
California 94122

Astronomical League
3939 Parkcrest Drive NE
Atlanta
Georgia 30319

Picture credits

The authors and publishers are grateful to the organizations and individuals who have granted permission to reproduce their photographs.

Abbreviations

AAT – Anglo-Australian Telescope Board; ESO – European Southern Observatory, FRG; NASA – National Aeronautics and Space Administration, USA; NOAO – National Optical Astronomy Observatories, USA; NRAL – Nuffield Radio Astronomy Laboratories, Jodrell Bank, UK; NRAO – National Radio Astronomy Observatory, USA; RGO – Royal Greenwich Observatory, UK; ROE – Royal Observatory, Edinburgh, UK.

Chapter 1 p. 8 John Fox/Manchester Evening News; p. 10 Bernard Lovell; p. 11 Francis Graham-Smith; p. 12 NRAO/AUI; p. 13 Bell Telephone Laboratories; pp. 14, 15 NRAL.

Chapter 2 p. 18 RGO/David Calvert.

Chapter 3 p. 28 Ann Ronan Picture Library and E.P. Goldschmidt & Co. Ltd.; p. 30 Science Museum, London; p. 31 The Royal Society, London; p. 33 RGO/David Calvert; p.34 Francis Graham-Smith; p. 35 NOAO; p. 36 (top) ROE; p. 37 (centre) AAT; p. 38 Multiple Mirror Telescope Observatory; p.39 NASA/Science Photo Library.

Chapter 4 p. 41 Mansell Collection; p. 42 the late Ronald Giovanelli; p. 43 University of Michigan; p. 44 ESO; p. 45 NRAL; p.46 Director, Arecibo Observatory, National Astronomy and Ionosphere Center; p. 47 NRAO; p. 48 Institute of Space and Astronautical Science, Tokyo; p. 49 NASA.

Chapter 5 p. 52 (top), p. 53 (top) Lowell Observatory; p. 53 (bottom), p. 56 Ann Ronan Picture Library; p. 59 Trevor Philip & Sons Ltd, (Specialists in early scientific instruments), 75a Jermyn Street, London SE1Y 6NP; p. 60 Science Museum, London.

Chapter 6 p. 62 Harold Hill; p. 64 Lick Observatory; p. 65 NASA; p. 67 NASA; p. 68 NASA.

Chapter 7 p. 72 NASA; p. 80 Encylopaedia Universalis, Paris.

Chapter 8 p. 84 NASA; p. 86 Print from the Soviet Union made available by Dr John Guest, University of London Observatory Planetary Image Centre; pp. 87, 88 (top R & L) NASA; p. 88 (bottom) US Geological Survey; p. 89 US Geological Survey/Alfred McEwen; p. 92 NASA.

Chapter 9 pp. 96, 97, 99, 100, 101 (top & centre) NASA.

Chapter 10 p. 104 Roy Panther; p. 105 Palomar Observatory; p. 108 Ann Ronan Picture Library; p. 109 ROE; p. 111 R.L. Waterfield and M.J. Hendrie; p. 112 Novosti Picture Agency; p. 113 ESA; p. 115 Max-Planck-Institut für Aeronomie, Lindau/Harz, FRG, courtesy Dr H.U. Keller.

Chapter 11 p. 116 James Shepherd; p. 117 NASA; p. 119 Steve Evans; p. 120 NOAO; p. 127 Horizons West, Arizona; p. 128 American Meteorite Laboratory.

Chapter 12 p. 129 NOAO; p. 132 (top) S. & J. Mitton; p. 133 (top) Serge Koutchmy; p. 133 (bottom) NASA; p. 135 RGO; p. 136 NASA; p. 139 Goran Scharmer.

Chapter 13 p. 144 AAT; p. 145 NASA; p. 148 ROE; p. 149 NOAO.

Chapter 14 p. 151 Yerkes Observatory; p. 156 Palomar Observatory © California Institute of Technology; p. 157 AAT; p. 158 NOAO; p. 160 NRAL.

Chapter 15 p. 163 ROE; p. 167 Lick Observatory.

Chapter 16 p. 170 (bottom L) NOAO; p. 170 (bottom R) NRAO/ AUI/S.P. Reynolds, R.A. Chevalier; p. 171 NRAO/AUI/R.J. Tuffs, R.A. Perley, M.T. Brown, S.F. Gull; p. 172 (top) University of Toronto/ Ian Shelton; p. 174 NOAO; p. 175 (top) NOAO.

Chapter 17 p. 178 Lund Observatory, Sweden; p. 179 Max-Planck-Institut für Radioastronomy, FRG; p. 180 Palomar Observatory; p. 183 AAT.

Chapter 18 p. 186 Akira Fujii; p. 187 (top) NOAO; p. 187 (bottom) S. & J. Mitton; p. 188 ESO; p. 189 NOAO; p. 190 AAT; p. 191 (bottom) ESO; p. 192 NOAO; p. 194 (top) ROE; p. 194 (centre) J. -L. Neito, Observatoires du Pic-du-Midi et de Toulouse; p. 195 NRAL; p. 196 NASA; p. 197 ROE.

Chapter 19 p. 200 Palomar Sky Survey; p. 201 NOAO; p. 203 NRAL; p. 204 Encyclopaedia Universalis, Paris, after K.I. Kellerman and I.K.I. Pauliny-Toth; p. 206 (top) NRAO/AUI/R.A. Perley, A.H. Bridle, A.G. Willis; p. 206 (bottom) NRAL; p. 207 NRAO/ AUI/E.B. Fomalont, R.A. Perley, A.H. Bridle, A.G. Willis; p. 209 NOAO.

Chapter 20 p. 213 University of Cambridge, Cavendish Laboratory; p. 215 Bell Telephone Laboratories.

Index